Lecture Notes in Computer Science 11474

Commenced Publication in 1973
Founding and Former Series Editors:
Gerhard Goos, Juris Hartmanis, and Jan van Leeuwen

More information about this series at http://www.springer.com/series/7409

Yukiko Kawai · Sabine Storandt ·
Kazutoshi Sumiya (Eds.)

Web and Wireless Geographical Information Systems

17th International Symposium, W2GIS 2019
Kyoto, Japan, May 16–17, 2019
Proceedings

Springer

Editors
Yukiko Kawai
Kyoto Sangyo University
Kyoto, Japan

Sabine Storandt
University of Konstanz
Konstanz, Germany

Kazutoshi Sumiya
Kwansei Gakuin University
Sanda, Japan

ISSN 0302-9743 ISSN 1611-3349 (electronic)
Lecture Notes in Computer Science
ISBN 978-3-030-17245-9 ISBN 978-3-030-17246-6 (eBook)
https://doi.org/10.1007/978-3-030-17246-6

LNCS Sublibrary: SL3 – Information Systems and Applications, incl. Internet/Web, and HCI

This Springer imprint is published by the registered company Springer Nature Switzerland AG
The registered company address is: Gewerbestrasse 11, 6330 Cham, Switzerland

Preface

These proceedings contain the papers selected for presentation at the 17th edition of the International Symposium on Web and Wireless Geographical Information Systems held in Kyoto in May 2019 and hosted by Kyoto Sangyo University.

The symposium is intended to provide an up-to-date review of advances in both theoretical and technical development of Web and Wireless Geographical Information Systems (W2GIS). The 2019 edition was the 17th in a series of successful events beginning with Kyoto 2001, and mainly alternating locations annually between East Asia and Europe. This was the third time the event was held in Kyoto, Japan. It provides an international forum for discussing advances in theoretical, technical, and practical issues in the field of wireless and internet technologies suited for the dissemination, usage, and processing of geo-referenced data.

W2GIS is organized as a full two-day symposium, recognized as a leading forum for the dissemination and discussion on the latest research and development achievements in the Web GIS and wireless domains. The submission process was successful this year, attracting 37 papers from almost all continents. This demonstrates not only the growing importance of this field for researchers but also the growing impact these developments have in the daily lives of all citizens. Each paper received three reviews and was ranked accordingly. The accepted 14 papers are all of excellent quality and cover topics such as web technologies and techniques, paths and navigation, web visualization, and novel applications. The proceedings also include a keynote abstract about five-dimensional world maps from the invited speaker Yasushi Kiyoki.

We wish to thank the authors who contributed to this workshop for the high quality of their papers and presentations and the support of Springer LNCS. We would also like to thank the Program Committee for the quality and timeliness of their evaluations. Finally, many thanks to the Steering Committee for providing continuous advice and recommendations.

February 2019

Kazutoshi Sumiya
Sabine Storandt
Yukiko Kawai

Organization

Program Committee

Masatoshi Arikawa	The University of Tokyo, Japan
Michela Bertolotto	University College Dublin, Ireland
Alain Bouju	La Rochelle University, France
James Carswell	Dublin Institute of Technology, Ireland
Christophe Claramunt	Naval Academy Research Institute
Sergio DiMartino	University of Naples Federico II, Italy
Zhixiang Fang	Wuhan University, China
Filomena Ferrucci	Università di Salerno, Italy
Jerome Gensel	Laboratoire d'Informatique de Grenoble, France
Ralf Hartmut Güting	Fernuniversität Hagen, Germany
Haosheng Huang	University of Zurich, Switzerland
Yoshiharu Ishikawa	Nagoya University, Japan
Farid Karimipour	University of Tehran, Iran
Yukiko Kawai	Kyoto Sangyo University, Japan
Kyoung-Sook Kim	Tokyo Institute of Technology, Japan
Daisuke Kitayama	Kogakuin University, Japan
Songnian Li	Ryerson University, Canada
Xiang Li	ECNU, China
Miguel R. Luaces	University of A Coruña, Spain
Miguel Mata	UPIITA-IPN, Mexico
Kostas Patroumpas	Athena Research Center, Greece
Sabine Storandt	University of Konstanz, Germany
Kazutoshi Sumiya	Kwansei Gakuin University, Japan
Taro Tezuka	University of Tsukuba, Japan
Hiroyuki Toda	NTT, Japan
Agnès Voisard	Freie Universität Berlin and Fraunhofer FOKUS, Germany
Yuanyuan Wang	Yamaguchi University, Japan
Yousuke Watanabe	Nagoya University, Japan
Robert Weibel	University of Zurich, Switzerland
Stephan Winter	The University of Melbourne, Australia
Junjun Yin	The Pennsylvania State University, USA
F. Javier Zarazaga-Soria	University of Zaragoza, Spain
Danielle Ziebelin	Université Grenoble-Alpes, France

"5-Dimensional World Map" A SPA-based Semantic Computing System for Global Knowledge-Sharing, Integration, Analysis and Visualization Towards Environmental Artificial Intelligence

Yasushi Kiyoki

Graduate School of Media and Governance, Faculty of Environment
and Information Studies, Keio University, SFC, 5322 Endo, Fujisawa,
Kanagawa, 252, Japan
kiyoki@sfc.keio.ac.jp
www.mdbl.sfc.keio.ac.jp
http://gesl.sfc.keio.ac.jp/

Abstract. Humankind, the dominant species on Earth, faces the most essential and indispensable mission; we must endeavor on a global scale to perpetually restore and improve our natural and social environments. One of the essential computations in environmental study is context-dependent semantic-computing to analyze the changes of various situations in a context dependent way with a large amount of environmental information resources.

The 5D World Map System and the Mathematical Model of Meaning technique (MMM) create new analytical system environments with the SPA concept (Sensing, Processing Analyzing and Actuation) and semantic computing. We have proposed the 5D World Map System, as "Global & Environmental Knowledge Creation and Sharing System", to collect and facilitate various social and natural-environmental information resources which are characteristics of natural phenomena, social infrastructure, disaster, traffic, water and energy supply.

The collaborative research activities are promising for approaching to the global architecture of a multi-visualized knowledge analysis and sharing system "5D World Map System" applied to "environmental semantic computing." The basic dimensional space of this system consists of a temporal (1st dimension), spatial (2nd, 3rd and 4th dimensions) and semantic dimensions (5th dimension), representing a large-scale and multiple-dimensional semantic space that is based on MMM.

This system is effective to make a progress of "Cyber-Physical integration" for detecting and analyzing natural and social environmental phenomena. The 5D World Map System is globally utilized as a Global Environmental Semantic Computing System, in SDGs 9, 11, 14, United-Nations-ESCAP: (https://sdghelpdesk.unescap.org/toolboxes). This dimensional space in 5D World Map memorizes and recalls various multimedia information resources (images, video, sound and text) with temporal, spatial and semantic correlation computing functions, and creates multiple views, applied to various "environmental information resources."

Keywords: Global environmental analysis · Semantic computing · Multimedia system · Data mining

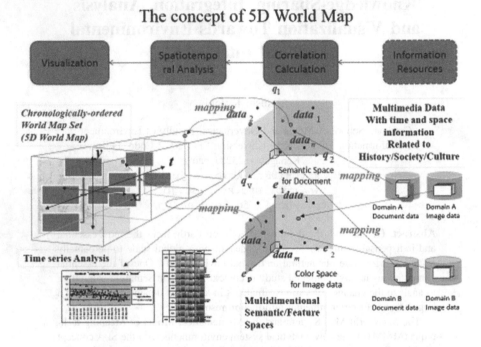

Fig.1. 5D World Map System for world-wide viewing for Global Environmental Analysis

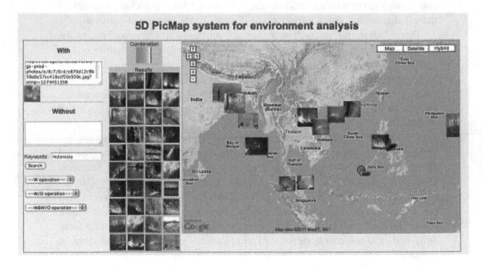

Fig. 2. Global Environmental Analysis of "Forest Fire" in "5D World Map System"

References

1. Kiyoki, Y., Kitagawa, T., Hayama, T.: A metadatabase system for semantic image search by a mathematical model of meaning. ACM SIGMOD Rec. **23**(4), 34–41 (1994)
2. Kiyoki, Y., Kitagawa, T., Hayama, T.: A metadatabase system for semantic image search by a mathematical model of meaning. In: Sheth, A., Klas, W. (eds.) Multimedia Data Management – Using Metadata to Integrate and Apply Digital Media–, McGrawHill (1998)
3. Kiyoki, Y., Ishihara, S.: A semantic search space integration method for meta-level knowledge acquisition from heterogeneous databases. Inf. Model. Knowl. Bases **14**, 86–103 (2002). IOS Press
4. Kiyoki, Y., Sasaki, S., Trang, N.N., Diep, N.T.N.: Cross-cultural multimedia computing with impression-based semantic spaces. In: Düsterhöft, A., Klettke, M., Schewe, K.D. (eds.) Conceptual Modelling and Its Theoretical Foundations, LNCS, vol. 7260, pp. 316–328. Springer, Heidelberg (2012). 10.1007/978-3-642-28279-9_21
5. Kiyoki, Y.: A Kansei: multimedia computing system for environmental analysis and cross-cultural communication. In: 7th IEEE International Conference on Semantic Computing, Keynote Speech, September 2013
6. Sasaki, S., Takahashi, Y., Kiyoki, Y.: The 4D world map system with semantic and spatiotemporal analyzers. Inf. Model. Knowl. Bases **XXI**, 18 (2010). IOS Press
7. Suhardijanto, T., Kiyoki, Y., Barakbah, A.R.: A term-based cross-cultural computing system for cultural semantics analysis with phonological-semantic vector spaces. Inf. Model. Knowl. Bases **XXIII**, 20–38 (2012). IOS Press
8. Kiyoki, Y., Chen, X., Sasaki, S., Koopipat, C.: Multi-dimensional semantic computing with spatial-temporal and semantic axes for multi-spectrum images in environment analysis. Inf. Model. Knowl. Bases **XXVI**, 20 (2016). IOS Press

References

Contents

An Iterative Two-Step Approach to Area Delineation

Xiang Li[1] , Qu Chen[1] , Buyang Cao[2(✉)] ,
Christophe Claramunt[3] , and Hong Yi[4]

[1] Key Laboratory of Geographic Information Science (Ministry of Education),
School of Geographic Sciences, East China Normal University,
Shanghai 200241, China
[2] College of Architecture and Urban Planning, Tongji University,
Shanghai 200092, China
caobuyang@tongji.edu.cn
[3] College of Logistics Engineering,
Shanghai Maritime University, Shanghai 201306, China
[4] Department of Real Estate, East China Normal University,
Shanghai 200241, China

Abstract. Recent advances of e-commerce development require the timely delivery of goods. Amongst many challenges to deal with, a logistics company should effectively delineate a service area for vehicles or persons to deliver goods or services to the clients with the minimal overall travel costs while balancing their workloads. Each service area contains a certain number of clients to be serviced, and the problem to be solved here is basically a spatial clustering one. However, most existing clustering methods usually ignore the objective of balancing workloads among clusters. This paper introduces an approach attempting to partition a service area effectively. The objectives of the problem include generating spatially continuous and mutually exclusive clusters (subareas), minimizing the travel distance, and balancing the workloads among clusters. A series of experiments are conducted in order to evaluate the performance of the proposed approach. Based on the benchmarks it appears that the proposed approach performs better with respect to the above three objectives.

Keywords: Clustering · Decision-making · Spatial analysis

1 Introduction

Recent advances of e-commerce development require timely deliveries. Logistics companies should efficiently delineate a specific delivery/service area for its vehicles or service persons in terms of balancing the workloads and minimizing overall travel distance. This problem can be classified as a Vehicle Routing Problem with Time Windows (VRPTW) according to Bräysy and Gendreau (2005). The strategy to solve this problem is clustering-first and routing-second. The clustering procedure here refers to divide the entire service area into multiple subareas. Without a loss of generality the

© Springer Nature Switzerland AG 2019
Y. Kawai et al. (Eds.): W2GIS 2019, LNCS 11474, pp. 1–12, 2019.
https://doi.org/10.1007/978-3-030-17246-6_1

clients in each subarea are served by one vehicle or service person, and the delivery schedule and path of the vehicle are generated by applying a given routing procedure.

In practice, each delivery vehicle must serve a group of clients who are clustered according to their geographical locations. To facilitate the business operations, the clusters and therefore subareas should be mutually exclusive and spatially compact. Furthermore, according to the business logic each subarea after the service area partition should have similar number or demands of clients in order to balance the workloads.

So far, a series of research efforts have been devoted to solve the above problem. Cao and Glover (2010) propose a clustering algorithm based on Thiessen-polygons and Tabu Search strategies trying to generate balanced and connected clusters. Özdamar and Demir (2012) introduce a hierarchical clustering algorithm to group demand points into clusters served by the corresponding vehicles to solve vehicle routing problems for disaster responses. With this methodology, the sizes of routing problems are reduced and can be solved in relatively short computational time that is crucial for disaster responses. In a similar context, Sheu (2007) utilizes a clustering algorithm to group demand points based on their priorities or similarities so that the emergency logistics operations can be effectively performed. The author applies real disaster data to demonstrate the potential of the proposed algorithm. In dealing with food supply chains, Bosona and Gebresenbet (2011) employ the clustering algorithm together with GIS techniques to divide the entire service area into several subareas where the collection centers are established. In this case, these collection centers are responsible for coordinating all delivery/pickup activities within the corresponding subareas. Based upon their experiments, travel times and distances as well as required vehicle routes are reduced significantly. Mesa-Arango and Ukkusuri (2015) also apply the clustering methodology for clustering lanes of demand in freight logistics networks.

Nowadays data-driven decision-making gets more popular, among which clustering algorithms attract a lot of attention. A relevant reference highlighting this role is provided in the work of Jain (2010). Cao et al. (2015) propose a Tabu Search based clustering algorithm that is capable of creating more cohesive clusters, which is crucial for many applications, e.g., classifications. Clustering algorithms or their variants are also applied to spatial data clustering problems. Zhang et al. (2007) generate geographic non-overlapping clusters, where geographic areas containing special features upon which some decisions are made. For instance, different administrative policies could be adapted for the subareas containing low-income households vs. the ones consisting of high-income households. She et al. (2017) for instance propose a Network-Max-P-Region (NMPR) model, in which people attempt to group n areas into the maximum number of regions until they meet a threshold. The authors also consider minimizing the heterogeneity while the influence of a street network is taken into account. Though the mathematical mode of NMPR is presented, it is an NP-hard problem and computationally intractable. Therefore, the authors propose a Tabu Search based heuristic to solve typical real-world problems. Certain randomly generated datasets and one dataset from the real application are employed to conduct the computational experiments. Nevertheless, to create the diversity of solutions, the proposed algorithm needs to generate a lot of random initial solutions that could eventually compromise the efficiency of the algorithm.

Overall, existing research projects on clustering algorithms emphasize on the metrics of similarities that can be represented by some distance measurements. However, to the best of our knowledge, such approaches hardly consider balanced clusters in terms of the workloads of the resultant clusters or subareas. Consequently, existing clustering algorithms do not balance the workloads among the different clusters. Albeit the algorithm proposed by Cao and Glover (2010) is able to handle balanced clusters to some degree, the algorithm is computationally expensive and unable to scale satisfactorily for larger datasets. This motivates us to seek a more efficient method that can create not only compact but also balanced clusters.

This study introduces a two-step approach whose objective is to solve the above mentioned clustering problem. The first step of the approach is to determine cluster centers, and the second one is to delineate the cluster boundaries through a shifting-insertion method in such a way that the resultant clusters are spatially continuous and have similar numbers/demands of clients.

The rest of the paper is organized in the following way. Section 2 defines the problem. Section 3 introduces the methodology, and Sect. 4 evaluates the performance of the proposed approach with a series of experiments. The paper conclude with some remarks.

2 Problem Definition

Let k denote the number of clusters or subareas to be created. Without losing generality let us assume one vehicle serves one subarea or cluster. Let N be a set of spatial objects standing for a group of clients with or without a specified demand. Both k and N are given. The clustering procedure is to assign every object in N to one of these k clusters reasonably.

Let $n_1, n_2, \ldots n_k$ denote the number of clients in each cluster. It is noted that the sum of set $n_1, n_2, \ldots n_k$ equals $|N|$. Let $c_1, c_2, \ldots c_k$ be the sum of client demands in each cluster, and $d_1, d_2, \ldots d_k$ represent the average distances in each cluster. The average distance of a cluster refers to the mean of distances between any two clients in the same cluster.

Furthermore, as given in Eqs. 1–5, let Std_n indicate the standard deviation of $\{n_1, n_2, \ldots n_k\}$, Std_c be the standard deviation of $\{c_1, c_2, \ldots c_k\}$, and Sum_d be the sum of $\{d_1, d_2, \ldots d_k\}$.

$$Std_n = \sqrt{\frac{1}{k}\sum\nolimits_{i=1}^{k}(n_i - \bar{n})^2} \tag{1}$$

$$\bar{n} = \frac{1}{k}\sum\nolimits_{i=1}^{k} n_i \tag{2}$$

$$Std_c = \sqrt{\frac{1}{k}\sum\nolimits_{i=1}^{k}(c_i - \bar{c})^2} \tag{3}$$

$$\bar{c} = \frac{1}{k} \sum_{i=1}^{k} c_i \tag{4}$$

$$Sum_d = \sum_{i=1}^{k} d_i \tag{5}$$

Using the above definitions, the objectives of the corresponding clustering problem are given as below.

- **Objective 1:** Clusters should be mutually exclusive in space and each cluster is spatially continuous.
- **Objective 2:** Minimize Std_n if clients have the same demand or do not have any specified demand, or minimize Std_c if clients have different demands.
- **Objective 3:** Minimize Sum_d.

The first objective not only facilitates delineating the delivery area for each vehicle but also tries to satisfy the operational requirements in real applications. The second one attempts to balance the workloads. When the demand of every client is the same or does not matter, minimizing Std_n is adequate. Otherwise, the objective is to minimize Std_c. The third objective ensures that each vehicle visits clients sequentially with the minimum expected travel distance.

As stated in Sect. 1, regular clustering algorithms target the first and third objectives to generate clusters as compact as possible. On the other hand, it is usually impossible to reach the all three objectives presented above simultaneously. Therefore, a novel approach to solve this problem is needed. In this paper we propose an approach based upon the *shifting-insertion* algorithm (Li et al. 2008) to provide an alternative to generate clusters upon these three objectives. The corresponding two-step approach is introduced in the next section.

3 Methodology

3.1 Principle

In order to apply the shifting-insertion (SI in short) algorithm, the center of each cluster needs to be determined first. We apply the maximum distances product (MDP in short) algorithm proposed by Xiong et al. (2011) to locate the cluster centers before SI algorithm is executed. Generally, the quality of a solution depends greatly on the locations of cluster centers, while the cluster centers derived from MDP algorithm may not guarantee a high quality solution. To address this issue, we propose an iterative procedure to find a new set of cluster centers based on the current resultant clusters and then apply SI algorithm to produce new clusters upon the new centers until the stop criterion (will be elaborated below) is satisfied. As a result, a number of solutions are generated and the best one with respect to the three objectives outlined above is returned as the final solution. Figure 1 illustrates the workflow of the proposed approach while the details are given in the following subsections.

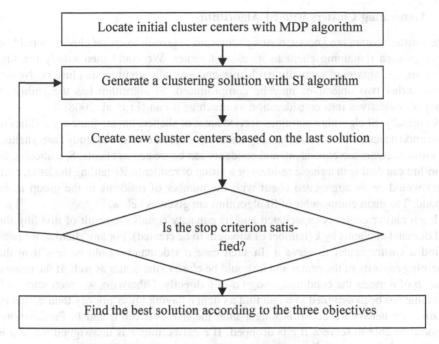

Fig. 1. An overview of the proposed approach.

3.2   Locating Cluster Centers with MDP Algorithm

The density-based MDP algorithm is implemented for selecting the initial cluster centers. Comparing to other common algorithms, it can produce results with less computational time, higher accuracy, and greater stability (Xiong et al. 2011). The objective of MDP algorithm is to locate centers in dense areas and these centers should be far away from each other.

Initially, each client itself is a candidate for certain cluster center. Let D_a be the search range of a given candidate α, and M_p be the number of clients covered by a circle with D_a as its radius and α as its center. D_a equals the mean of distances between all pairs of clients in the underlying dataset. For any candidate, if its M_p is larger than a user-defined threshold, the client is considered as a high density one. All high density clients are stored in a sorted set D and the highest density client from D is selected as the first center C_1. At the same time, C_1 is removed from set D. Next, the farthest high density client from C_1 among all remaining points in D is chosen as the second center C_2. The procedure then continues as follows: For each remaining client x_i in set D, the distances from x_i to all selected centers are calculated and defined as $d(x_i, C_1), d(x_i, C_2), \ldots, d(x_i, C_j)$, where j is the number of selected centers. Let P_i be the product of $d(x_i, C_1), d(x_i, C_2), \ldots, d(x_i, C_j)$. We pick x_i with the maximum P_i as the next center C_{j+1}. Repeat the above steps until k centers are selected.

3.3   Generating Clusters with SI Algorithm

If the cluster centers are known, a straightforward approach to create clusters would be to assign each remaining client to its closest center. We could then satisfy the first objective, i.e. obtaining mutually exclusive and spatially continuous clusters. Nevertheless, other two objectives may be compromised. SI algorithm has the ability of taking all objectives into consideration as much as it can (Li et al. 2008).

Originally, SI algorithm partitions service areas of shelters having finite capabilities to accommodate residents who are spatially unevenly distributed. For our study case, shelters and cluster centers are equivalent, and residents can be treated as clients. Specifically, SI algorithm can deal with a single resident or a group of residents. Regarding the latter, each group would be an aggregated client with the number of residents in the group as its demand. The main components of SI algorithm are given as below.

Each cluster center is capacitated and its capacity equals the result of dividing the total demand of clients by k (number of subareas to be created). For any client a, we need to find a cluster center to serve it. In such case a's demand should be less than the remaining capacity of the center and it should be close to the center as well. If the nearest center b of a meets the condition, assign a to b directly. Otherwise, we need scan each client that has been assigned to b and find a client c having lower priority than a, that is, the distance between c and b is longer than the one between a and b. Furthermore, b should be able to serve a if c is dropped. If c exists, then c is unassigned while a is assigned to b. If no such client exists, a may have to be assigned to another center with enough spare capacity. However, in this situation, spatial continuity may be violated. To resolve this issue, we need to find a *shifting client* that can be switched with a without violating spatial continuity and minimize the incremental travel distance. The procedure repeats until all clients have been assigned to cluster centers.

3.4   Solution Procedure

The two steps discussed previously are integrated into the following procedure for the purpose of partitioning an area reasonably.

a. Finding all initial cluster centers with MDP algorithm
b. Generating the initial clusters with SI algorithm
c. While the stop criterion is not satisfied
   (1) For each cluster $i = 1, \ldots, k$ in the previous solution, do
      i. Calculate the geometric center nc of clients in cluster i
      ii. Set nc to be the new cluster center for cluster i
   (2) End for
   (3) Updating clusters with SI algorithm based on the k new cluster centers
d. End while
e. Return the best solution as the final output

In the above procedure, MDP algorithm is used once for creating the initial cluster centers while SI algorithm is applied to generate the initial clusters. During the iterations, the new cluster centers are found upon the previous solution and SI algorithm generates the new clusters. By applying SI algorithm, all three objectives can be satisfied to a great extent. In order to evaluate the quality of solutions, different metrics can be adopted with respect to the different preferences. If a logistic company desires the resultant clusters to be as compact as possible to reduce travel cost, the solution with the least Sum_d would be the most suitable one. If one wishes each delivery unit (e.g. a vehicle, a service person, etc.) could have a balanced workload, then the solution with the least Std_n or Std_c would be the best one. The latter is usually more popular in the logistic applications.

The stop criterion is defined as follows. The maximum number of iterations could be defined. Nevertheless, it impacts the performance (solution quality and computational time) of the algorithm. It is conceivable that a large iteration number may help secure a high-quality solution but require longer computational time. As a trade-off we may add an additional stop criterion. For certain applications, it may be possible to set an expected range of Sum_d, Std_n, or Std_c, for example $Std_n < 1$. At times, once the number of iterations reaches the maximum or the metrics (Sum_d, Std_n, or Std_c) of the current solution falls in the expected range, the solution procedure terminates.

4 Computational Experiments

4.1 Data Preparation

The practical dataset from a real logistics application is employed to evaluate the performance of the proposed algorithm. The delivery or service area is located in the downtown area of Nanning, Guangxi Zhuang Autonomous Region, China. 2216 clients are distributed in this area. They are authorized vendors of some special products and need delivery services. According to the requirements of the application, these clients need to be grouped into five clusters for daily delivery services. The number of clients in each cluster should be as similar as possible, and the clusters need to be spatially as compact as possible. Figure 2 illustrates the distribution of clients.

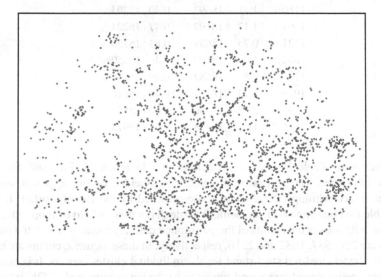

Fig. 2. Distribution of clients in the dataset

The clients are spatially unevenly distributed and each of them has the same demand. For the benchmarking purpose, the algorithm proposed by Cao and Glover (2010) (BC algorithm) is implemented and applied.

The proposed approach in this paper is implemented using C# and is carried out on a desktop with 3.3 GHZ CPU and 8 GB memory. We evaluate the solution quality in terms of Std_n or Std_c. In the following experiments, we do not set an expected range for stopping criterion, but only define a maximal iteration number (20).

4.2   Benchmark

In order to evaluate the scalability of the proposed approach, 12 groups of clients are randomly selected from this dataset to form sub-problems. The number of clients in each sub-problem varies from 250 to 2216. Both BC algorithm and the proposed approach are applied to these sub-problems, respectively. Since the clients have the same demand for this dataset, we use Std_n to judge the quality of solutions. After all iterations are performed, the solution with the lowest Std_n will be selected as the final solution yielded by the proposed approach. Table 1 presents the solutions obtained by the proposed approach and BC algorithm.

Table 1. Comparing solutions with BC algorithm and the proposed approach

#Clients	BC algorithm		The proposed approach	
	Std_n	Sum_d (m)	Std_n	Sum_d (m)
250	0.63	16359	0.40	15119
534	1.6	16168	0.49	15571
767	4.5	17874	0.75	16135
987	1.02	18715	0.40	15994
1180	1.41	18692	0.40	15703
1361	1.17	18197	0.47	16224
1521	0.75	18637	0.63	15755
1682	13	18603	0.40	15650
1798	0.8	18587	0.49	15636
1945	5.73	18187	0.40	15816
2081	27.84	18386	0.00	15559
2216	12.35	18338	0.63	15973

According to Table 1, one can conclude that in terms of Std_n and Sum_d, our approach dominates. The values of Std_n and Sum_d yielded by our approach are more stable and overwhelmingly lower than those obtained by the BC algorithm. It is also conceivable that our approach can achieve **Objectives two** and **three** more effectively.

Figure 3 illustrates the results of the partitioned delivery areas in which the numbers of clients are 250, 987, 1682 and 2216, respectively. In these pictures, clients are colored upon their clusters and red stars stand for the individual cluster centers. It is noted that clusters are spatially continuous and mutually exclusive as expected in **Objective one**.

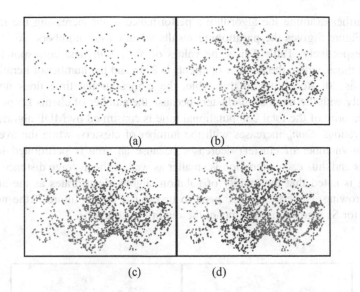

(a)   (b)

(c)   (d)

Fig. 3. Results of the partitioned delivery areas when the numbers of clients are 250 (a), 987 (b), 1682 (c) and 2216 (d), respectively. (Color figure online)

Figure 4 depicts the values of Std_n during the solution procedure for the numbers of clients are 250, 987, 1682, and 2216 respectively. The maximal number of iterations of the solution procedure is 20. One can notice that a smaller Std_n usually appears after a few iterations.

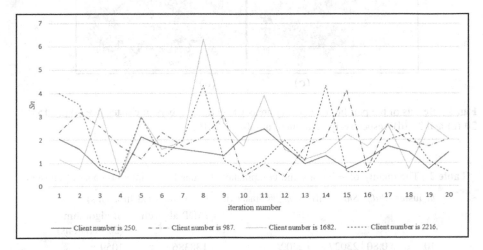

Fig. 4. Changes of Std_n during the procedure where the numbers of clients are 250 (a), 987 (b), 1682 (c) and 2216 (d), respectively.

We further examine the algorithm's performance when increasing the number of clusters. Figure 5 gives the partitioning results when cluster numbers are 10, 15, 20, and 25, respectively. Table 2 lists the values of Std_n, Sum_d and corresponding computational times, where the number of clients is 2216 and the number of iterations of SI algorithm is 20. According to this table, the computational time does not change significantly with the number of clusters because the number of clients keeps the same and the majority of the total computational time is consumed by MDP algorithm upon our observations. Sum_d increases with the number of clusters, while the average distance (Sum_d/number of cluster) decreases because an area is partitioned into more subregions and thus each subregion is smaller as well as the average distance. In terms of Std_n, it is noted that the quality of solution slightly deteriorates as the number of clusters growing. Nevertheless we need to investigate in the future if the number of iterations for SI algorithm will impact the quality of solutions.

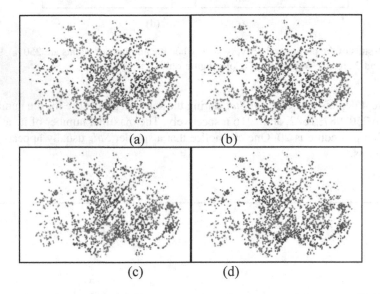

(a)   (b)

(c)   (d)

Fig. 5. Results of the partitioned delivery areas where the numbers of clusters are 10 (a), 15 (b), 20 (c) and 25 (d), respectively, and client number is 2216.

Table 2. The quality of solutions and computational times for various numbers of clusters

#Cluster	Std_n	Sum_d (m)	Sum_d/#Cluster (m)	Computational time (ms)	
				MDP algorithm	SI algorithm (20 iterations)
10	0.80	23027	2302	138386	1059
15	1.88	27328	1821	138041	826
20	1.99	30457	1522	137529	1026
25	2.61	35901	1436	137373	971

5 Conclusions

This paper introduces an approach whose objective is to partition a given geographic area into several spatially continuous groups or subareas. The peculiarity of our algorithm is that it generates spatially continuous and mutually exclusive clusters, minimizing travel distance, and balancing workloads among clusters. The approach consists of two steps, namely, generating initial cluster centers by using MDP algorithm and creating clusters by applying SI algorithm iteratively. We believe that the principles developed by our approach are applicable for many logistic applications.

The computational experiments have been conducted to evaluate the performance of the proposed approach. The data from practice has been collected and used for the computational experiment, while an existing algorithm is employed for the benchmark evaluation. The results show that our approach can yield better solutions comparing to the benchmarked algorithm.

Even though our approach is able to achieve satisfactory solutions, how to set a reasonable stop criterion for the algorithm remains an open question. Future work should also focus on the efficient creation of initial cluster centers as this step is currently the most time-consuming part of the proposed algorithm.

Acknowledgements. We are indebted to three anonymous reviewers for insightful observations and suggestions that have helped to improve our paper. The work is partially supported by the projects funded by National Natural Science Foundation of China (grant numbers: 41771410 and 41401173).

References

Barreto, S., Ferreira, C., Paixao, J., Santos, B.S.: Using clustering analysis in a capacitated location-routing problem. Eur. J. Oper. Res. **179**, 968–977 (2007)

Bosona, T.G., Gebresenbet, G.: Cluster building and logistics network integration of local food supply chain. Biosys. Eng. **108**, 293–302 (2011)

Bräysy, O., Gendreau, M.: Vehicle routing problem with time windows, part I: routing construction and local search algorithms. Transp. Sci. **39**, 104–118 (2005)

Cao, B., Glover, F.: Creating balanced and connected clusters for improved service delivery routes in logistics planning. J. Syst. Sci. Syst. Eng. **19**, 453–480 (2010)

Cao, B., Glover, F., Rego, C.: A tabu search algorithm for cohesive clustering problems. J. Heuristics **21**, 457–477 (2015)

Jain, A.K.: Data clustering: 50 years beyond K-means. Pattern Recogn. Lett. **31**, 651–666 (2010)

Li, X., Claramunt, C., Kung, H.T., Guo, Z.Y., Wu, J.P.: A decentralized and continuity-based algorithm for delineating capacitated shelters' service areas. Environ. Plan. B: Plan. Des. **35**, 593–608 (2008)

Mesa-Arango, R., Ukkusuri, S.V.: Demand clustering in freight logistics networks. Transp. Sci. Part E **81**, 36–51 (2015)

Özdamar, L., Demir, O.: A hierarchical clustering and routing procedure for large scale disaster relief logistics planning. Transp. Sci. Part E **48**, 591–602 (2012)

She, B., Duque, J., Ye, X.: The Network-Max-P-Regions model. Int. J. Geogr. Inf. Sci. **31**, 962–981 (2017)

Sheu, J.B.: An emergency logistics distribution approach for quick response to urgent relief demand in disasters. Transp. Sci. Part E **43**, 687–709 (2007)

Xiong, Z., Chen, R.T., Zhang Y.F.: Effective method for cluster centers' initialization in K-means clustering. Appl. Res. Comput. (2011)

Zhang, B., Yin, W.J., Xie, M., Dong, J.: Geo-spatial clustering with non-spatial attributes and geographic non-overlapping constraint: a penalized spatial distance measure. In: Zhou, Z.-H., Li, H., Yang, Q. (eds.) PAKDD 2007. LNCS (LNAI), vol. 4426, pp. 1072–1079. Springer, Heidelberg (2007). https://doi.org/10.1007/978-3-540-71701-0_121

City Link: Finding Similar Areas in Two Cities Using Twitter Data

Wannita Takerngsaksiri[1](✉), Shoko Wakamiya[2], and Eiji Aramaki[2]

[1] Chulalongkorn University, 254 Phayathai Road, Pathumwan, Bangkok, Thailand
wannita.tkn@gmail.com
[2] NAIST, 8916-5 Takayama-cho, Ikoma, Nara 630-0192, Japan
{wakamiya,aramaki}@is.naist.jp

Abstract. Today in our increasingly globalized world, the number of people travelling overseas is increasing. A system that helps overseas travelers by providing information related to unfamiliar places has been earnestly sought. This study develops such a system by exploiting user-generated data over a popular social network platform: Twitter. We propose the use of natural language processing (NLP) as a method of estimating location similarity between areas in different cities. Finally, location similarity is visualized on a map. Our experiment is conducted at two popular sightseeing cities: Bangkok, Thailand and Kyoto, Japan. Our evaluation using crowd-sourcing-based 1,000 questionnaires empirically demonstrated that the proposed method can find similar places in the two cities. This result demonstrated the fundamental feasibility of our approach.

Keywords: City similarity · Crowd-sourcing · Doc2Vec · Map visualization · Social media

1 Introduction

Year by year, the number of people visiting overseas has increased dramatically. For instance, worldwide, more than 1.2 billion people traveled in 2016[1]. According to Japan National Tourism Organization (JNTO) data, overseas visitors to Japan have increased sharply, reaching 28 million people in 2017, 4 million more than the prior year[2]. At the same time, this sudden increase raises the issue that sufficient information services for foreigners have not been provided. Consequently, many foreign people must confront difficulties when forming their travel plans because of lack of knowledge related to unfamiliar cities.

Various information sources are already available. A popular source is the internet. Many public and private location reviews have been shared over the

---

[1] https://data.worldbank.org/indicator/ST.INT.ARVL?end=2016&start=1995&type=shaded&view=chart.

[2] https://www.tourism.jp/en/tourism-database/stats/inbound/.

© Springer Nature Switzerland AG 2019
Y. Kawai et al. (Eds.): W2GIS 2019, LNCS 11474, pp. 13–27, 2019.
https://doi.org/10.1007/978-3-030-17246-6_2

internet. Nevertheless, some are unreliable[3] (too old, fake information, just a rumor). More importantly, most reviews are not comprehensible by everyone. An example is the following review describing

"*Ginkaku temple*" in Kyoto[4]: *Ginkaku temple has a calm stance to feel wabisabi. I prefer this temple to Kinkaku temple.*

This review, because it is vague and one must guess the degree of calmness of the place, is not informative for people who have never been to Japan or who do not know about Japan well. By contrast, comparative information based on familiar places would be helpful to find and learn about unfamiliar places. For example,

Ginkaku is similar to Ayutthaya ruins, but very crowded like Siam

would make it easier for tourists from Thailand to understand the atmosphere of Ginkaku temple. However, it would be difficult to present this type of information because non-Thai people or people who have never been to these places can not understand a comparison between Thailand and Japan. To meet the common needs of foreign people for other countries than the two above, a (semi-)automatic method is necessary to create comparisons that include cities in a given pair of countries.

Our solution is social media based on two cities' comparisons. Today, people using social media are increasingly expressing their activities, feelings, and stories with full emotions. Especially, earlier reports have described that Twitter has over 6,000 new tweets updated every second on average[5], producing 500 million tweets per day and around 200 billion per year. We regard such a vast amount of data as a valuable data source to inform us about stories behind places. For example, places that have many checkin tweets tend to be sightseeing places and those where people talking about food might be restaurants.

The basic work flow is the following: First, we extract Twitter-based city data independently using Twitter data. We then estimate the similarity of areas in different cities such as between a target city and a visitor's hometown using tweet data. For similarity calculation, we use Doc2Vec and cosine similarity. Finally, we visualize similar area correspondence using a Google Maps API. This paper describes a case study of automatic city comparison between Kyoto, Japan and Bangkok, Thailand.

2 Related Work

Because one city is definable both objectively and subjectively, numerous studies have attempted to describe the atmosphere of the city behind the concrete map.

---

[3] https://www.businessinsider.com/20-percent-of-yelp-reviews-fake-2013-9.

[4] http://kyoto-kanko.net/kyoto-spot/kyoto-spot-847/.

[5] http://www.internetlivestats.com/twitter-statistics/.

The first attempt is "The Image of the City" by Lynch [1], representing cities using five fundamental elements: path, edge, district, node, and landmark. The results were obtained via interviews to residents of the cities in the United States of America.

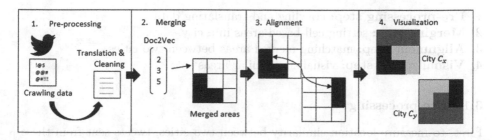

Fig. 1. Flow of the proposed method. The method comprises four steps. (1) Pre-processing, (2) Merging, (3) Linking, and (4) Visualization.

Subsequently, Nold created the "Emotion Map" [2] in Greenwich using a special device to detect galvanic skin response (GSR), which can indicate a wearer's emotional state the results were combined together with the geographical location.

Aside from data related to workshops, surveys, and interviews, social media data have been available. Many studies of map visualization have used social media data to extract specific functions in cities. Quercia et al. used social media data such as geo-referenced picture tags from Flickr and Instagram and geo-referenced tweets from Twitter to create a map called "Happy Map" [3] in London and Boston and "Smelly Map" [4] in London and Barcelona. Aiello et al. created a "Chatty Map" [5] in London and Barcelona using similar social media data.

We are interested in Twitter data because many earlier studies have indicated that Twitter data can tell us many stories, for example, related to health [6], political [7,9] and crime [8]. Hahmann et al. demonstrated that Twitter data contexts have sometimes been related to locations [10]. Therefore, we aim to insights into similarity between cities by exploiting Twitter data. Some earlier studies have found similar cities using Foursquare [11] and query-logs [12,13]. For the Foursquare-based approach, Preotiuc-Pietro et al. presented similar cities based on similar venues in cities across the United States. Research using query-logs can show the interests of the city's inhabitants. However, both Foursquare and query-log data are mostly emotionless. Consequently, our research uses Twitter, with which people apparently express their feelings more.

3   Method

This study proposes a method to find similar areas between cities. The method comprises four steps listed as follows. Figure 1 presents the basic flow of the proposed method.

1. **Pre-processing step:** crawling and translating data
2. **Merging step:** setting cell-based areas in a city
3. **Alignment step:** matching merged areas between two cities
4. **Visualization step:** visualizing similar areas

3.1   Pre-processing

First, to measure location similarity between two cities, tweets sent from these cities are required.

A Twitter dataset is collected through the Twitter real-time streaming API[6]. The API provides filtering of two types to collect tweets: (1) language filtering and (2) location filtering as shown below.

1. **Language filtering:** We use only tweets written in the official language of a target city and discard tweets written in others. Let two cities be Bangkok, Thailand, and Kyoto, Japan, tweets written in Thai and those written in Japanese be extracted respectively as Bangkok dataset, and Kyoto dataset. Then, to address the two language texts in the same method, tweets are translated into a common language (such as English). Therefore, we apply the Google translate API from the library[7] to the tweet texts and locations. However, the accuracy of the translation depends on the languages translated. According to Google's own tests[8], the Chinese to English translation received a rating of 4.3 out of 6. Thai and Japanese could receive even lower scores because of their complexity.
2. **Location filtering:** Our method requires tweets with location information. Location information of two types is used as follows.
   - GPS (i.e., [longitude : latitude, 13.4444 : 43.2222])
   - Bounding box (represented using a place name i.e., "*Siam City*")
   A bounding box is used to apply a scope of study to a target city area. This study sets the size of the bounding box to be in approximately 10 km in both height and width. Then it is located to cover the target city's central area and popular sightseeing places. For example, a bounding box for Bangkok, Thailand covers Siam, Sukhumvit, and The Grand Palace and that for Kyoto, Japan covers Kyoto Station, Gion-Shijo, and The Imperial Palace. Subsequently, we extract only those tweets with the GPS location using their coordinates attribute.

---

[6] https://developer.twitter.com/en/docs/tweets/filter-realtime/overview.
[7] https://github.com/mouuff/mtranslate.
[8] https://www.argotrans.com/blog/accurate-google-translate-2018/.

Moreover, we cleaned the collected tweet texts by replacing part of the URL with a tag <URL>, the whole checkin tweets via Location Based Social Network (LBSN) such as Foursquare with the tag <CHECK IN>, and the described username in reply tweets with the tags <MENTION>, respectively, using Regex. We also eliminated all emoticons from tweet texts.

3.2   Area Merging

To align target cities by finding similar areas in terms of tweets, we define areas in a city. Therefore we initially divide the bounding box, which was set for collecting a city's tweet dataset, into a grid location system. The bounding box is divided into 225 (=15 × 15) equal parts vertically and horizontally, according to the demarcation method of the basic grid square[9], which divides a partition into about 1 km on a side. We regard the grid cell as an area. Therefore, each city is split into 225 areas.

Although a city is divided into small areas of equal size, some areas having similar characteristics should be merged together and be regarded as one area. Therefore, similarity between areas is measured based on tweets in each area. Given areas a_i and a_j, tweets in these areas T_i and T_j are regarded as documents D_i and D_j by concatenating the tweets. The documents are used to train on Gensim[10] Distributed memory Doc2Vec to receive their document vectors, d_i and d_j. To validate how many epochs and dimensions should be used for Doc2Vec, we maximize the number of documents for which similarity with themselves ranks first.

The similarity between areas, a_i and a_j, is measured using the cosine similarity function Cos of the document vectors d_i and d_j as shown below.

$$Sim(a_i, a_j) = \frac{Cos(d_i, d_j) + Cos(d_j, d_i)}{2 * (|T_i| + |T_j|)} \qquad (1)$$

Because similarity between two areas is not equal bi-directionally, we consider both directions' similarities in Eq. 1.

Furthermore, we found that some unique areas were not merged or were merged with all others easily in preliminary experiments. Therefore, we also use the number of tweets in the areas, $|T_i|$ and $|T_j|$, in Eq. 1. For illustration, an area that has no tweets at all tends to be merged only with a neighbor area that has no tweets either and which was never merged with others, although an area that has numerous tweets is likely to be merged with any neighbor area easily. However, if both areas have no tweets, then they are merged together to prevent a number divided by 0.

Merging Two Single Areas. First, we find similar areas based on similarity sim between proximate areas as

$$MA_x = \{(a_i, a_j) | Sim(a_i, a_j) > \alpha \wedge Neighbor(a_i, a_j) = 1 \wedge a_i, a_j \in A_x\} \qquad (2)$$

---

[9] http://www.stat.go.jp/english/data/mesh/05.html.
[10] https://github.com/RaRe-Technologies/gensim.

where Sim represents the similarity function defined in Eq. 1, $Neighbor$ stands for the function to return 1 if two areas are neighbors, A_x denotes a set of areas in a city c_x, MA_x represents a set of merged areas in the city c_x, and α expresses a threshold.

The algorithm loops all areas and finds similarity between each area and its neighbors if the similarity is more than the threshold then the two areas are mutually merged.

After looping all areas (225 areas), the overall leftover areas are checked continuously with a lower threshold than the initial loop and are merged while the number is still more than a stop condition to exit a loop. The maximum numbers of areas in a merged area are limited considering an area size limitation.

Merging Multiple Areas. The algorithm to merge multiple areas such as a merged area and a single area or another merged area is somewhat different from the above. Because the documents are not merged together during merging of two single areas, we cannot find the document vector of merged areas. Therefore, we find a representative similarity for a merged area. By calculating the similarities between areas in merged areas of two cities, the most similar pair is stated as the representative similarity of the two merged areas. The similarity between an area a_i and its neighboring merged area ma_a would be calculated as shown below.

$$Sim'(a_i, ma_a) = \arg \max(Sim(a_i, a_j \in ma_a)) \qquad (3)$$

Additionally, the similarity between a merged area ma_a and its neighboring merged area ma_b would be calculated as follows.

$$Sim'(ma_a, ma_b) = \arg \max(Sim(a_i \in ma_a, a_j \in ma_b)) \qquad (4)$$

3.3   Linking Areas

After the area-merging step above, we link areas between two cities as c_x and c_y. We start by merging all documents in each merged area together. Each document now comprises tweets in a merged area instead of an original area. Then, we use Gensim Doc2Vec to train them in the same way as the merging algorithm and compute the cosine similarity. We sorted the similarity for each merged area in descending order to create a ranking called $MostSim$, which would rank only the merged areas in the other city. For example, $MostSim_{ix}$ of a merged area ma_i in c_x would have only the list of the merged areas in c_y, and vice versa, as

$$MostSim_{ix} = Sort(SimList_{ix})$$
$$SimList_{ix} = \{(ma_j, Sim(ma_i, ma_j)) | ma_j \in MA_y\},$$

where $SimList$ comprises pairs of a merged area in the other city and its similarity with a given merged area and where the function $Sort$ is to sort a given list based on the similarity in descending order.

However, a duplication problem occurs by which the same areas can be picked up as the most similar areas with several areas. To address this difficulty, we provide two algorithms: (1) greedy algorithm and (2) queue algorithm.

Greedy Algorithm. The greedy algorithm links the most similar merged areas together. If findings show that the most similar one or the previous one is already used, then the next most similar merged area is selected immediately. The greedy algorithm is described in Algorithm 1.

Algorithm 1. Greedy Algorithm

1: **for** x in merged area **do**
2:     **if** $MostSim(x)$ is used **then**
3:         pick up next $MostSim(x)$
4:     **else**
5:         assign $MostSim(x)$
6:     **end if**
7: **end for**

Queue Algorithm. The greedy algorithm makes a link between cities without waiting, often causing a local minimum. However, to search all combinations optimally requires a heavy calculation. Therefore, we propose a queue algorithm as expressed below: if the first rank of $MostSim$ of a merged area has already been used, then the queue algorithm pushes the merged area back into a queue to wait for the next iterator of selecting the coming up rank. Therefore, the queue algorithm more highly prioritizes on the first rank overall. The queue algorithm is described in Algorithm 2.

Algorithm 2. Queue Algorithm

---
    waiting queue ← merged area
2: $i \leftarrow 1$
    while size(waiting queue)> 0 **do**
4:     **for** x in merged area **do**
        if $MostSim(x)[i]$ is used **then**
6:             put x in temp
        else
8:             assign $MostSim(x)$
        end if
10:    **end for**
    waiting queue ← temp
12:     $i+ = 1$
    end while

In addition, to ascertain which algorithm is suitable for linking cities, we apply two criteria: the sum of overall similarity after linking and the maximum of minimum pair similarity. In this study, although we prioritize more on the sum of overall similarity, it would be better if both criteria reached the maximum point.

3.4 Visualization

Map Visualization. The link between two cities based on similar areas is visualized on a map. Merged areas that are linked to be the most similar to each other would be the same colors. Moreover, to distinguish the merged areas clearly, other than colors, we also use a pin with an English alphabet to pair merged areas. A conceptual map visualization is shown in Fig. 2.

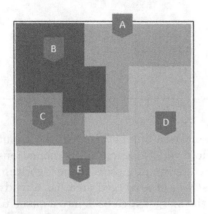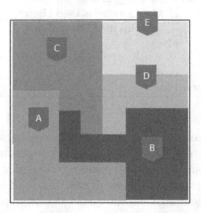

Fig. 2. Conceptual image when two cities are mutually aligned. (Color figure online)

Table 1. Points of bounding box in Bangkok and Kyoto

City	Southwest point	Northeast point
Bangkok	100.484605, 13.685780	100.584605, 13.815780
Kyoto	135.675300, 34.955205	135.795300, 35.055205

4 Experiment

4.1 Settings

Twitter Data (Bangkok and Kyoto). For this experiment, we chose Bangkok and Kyoto as our target cities. We analyzed tweets from these two cities to ascertain similar areas. The languages we used were English and Thai for the Bangkok dataset and English and Japanese for the Kyoto dataset. The area of the bounding boxes is shown in Table 1.

The tweets were collected for two months from 1 June through 31 July 2018. Table 2 presents the number of tweets in respective cities; Table 3 presents sample tweets.

Table 2. Statistics of the tweet dataset

City	# of tweets
Bangkok	60321
Kyoto	74742

After collecting all tweets and separating them into grid areas, we used 80 vector dimensions with approximately 35–50 epochs to train on Doc2Vec in Gensim. The values of these parameters were set because they maximized the validation number (Sect. 3.2) properly.

Table 3. Samples of tweet datasets. The first three tweets from the top are of the Bangkok dataset. The others are of the Kyoto dataset.

Original Tweet	Translated Tweet	Location
เพื่อนบอกฟู้ดคอร์ทนั้น ไม่เห็นอร่อยเลย	Friends tell the food court. Do not see the delicious.	Si Phraya
เหนื่อย พลังลดไวเกิน เมาแดด	Tired of the sun	Ratchathewi
เรายังตื่นเต้นขนาดนี้ แล้วพิเฌอจะขนาดไหนวะเนี่ยย สู้ๆนะพิเฌอ #CherPrangBNK48	We are also excited about this. I will not be able to do that. #CherPrangBNK48	Silom
I'm at わかさスタジアム京都 in Kyoto, 京都府 https://t.co/34oXyqq1le	<CHECK IN>	Kami-ku
ここで勝てないと B クラスが現実味を帯びてしまう、だもんで最低 2 勝を	If you can not win here, B class will take on reality, let's have at least 2 wins	Fushimi-ku, Kyoto-shi
ビールうまい。このあとレコ祭行ききます。	Beer good. I will go to the festival after this.	Kyoto City, Nakagyo-ku

For merging conditions, we used the initial threshold α as 0.7 and the stop condition as 12. The area size limitation, which indicates the maximum number of areas in a merged area, was set as 25 because the ideal size is 225 areas divided by 12, which is 20. However, because making all merged areas reach the ideal size at once is difficult, we set it somewhat higher. The rate of decrease of threshold α also affected the result. The value should be small to merge areas gradually, although the smaller value would require more time to compute. Therefore, we used 0.0001 as the initial value and decreased it when the number of merged areas almost reached the stop condition.

For linking up cities, we used the queue algorithm as that time during which the sum of the overall similarity was higher than the queue algorithm.

Gold Standard Data (Amazon Mechanical Turk Data). It is difficult to validate our map because no one knows what result is the best or the correct one.

Therefore, to validate the results, we decided to gather people's opinions based on their experience through a crowd-sourcing service. Amazon Mechanical Turk[11] was chosen as our crowd-sourcing platform because of its worldwide reputation. Finally, we were able to collect answers of the most similar pairs of places in the two cities from people around the world.

Detailed steps of crowd-sourcing are the following. First, we had six pre-qualification questions to check the participants' reliability and knowledge related to the cities, which separated into two types. One would ask participants to match the place pictures with the names of the places, whereas another asked about the name of a river in the cities: Bangkok and Kyoto. Then we eliminated all answers of participants who answered incorrectly on the six pre-qualification questions. Therefore, we collected 509 assignments at first and 387 assignments after elimination. In the question set, we used 12 places in each city, which were chosen from popular places around the target areas. Table 4 lists 12 places in Bangkok and in Kyoto. For places in the cities, the participants answered the most similar places in the other city, and vice versa.

Table 4. Candidate places for the Amazon Mechanical Turk survey.

Cities	Places
Bangkok	Asiatique, Asok, Chatuchak Market, Chinatown, Chulalongkorn University, Golden Mountain Temple, Siam, Silom, The Grand Palace, Victory Monument, Wat Arun Ratchawaram, Wongwian Yai
Kyoto	Arashiyama, Fushimi-Inari, Gion-Shijo, Katsura, Kinkaku-ji, Kiyomizu-dera, Kyoto Imperial Palace, Kyoto Station, Kyoto University, Nijo Castle, Nishiki Market, Yasaka Shrine

In the validation part, the answers would be sorted to make a majority ranking. This ranking was used for comparison with the map generated using the proposed method. The percentages of the answers from crowd-sourcing are shown in Figs. 3 and 4. Figure 3 shows the relation from the places in Kyoto to the ones in Bangkok, whereas Fig. 4 shows the relation from the places in Bangkok to those in Kyoto.

4.2   Visualization

Doc2Vec Matrix Visualization. To validate the Doc2Vec, the most similar area should be itself. Figure 5 is the visualization of the cosine similarity of each area. In both axes, area numbers from 0 to 11 refer to the merged areas in Bangkok, whereas area numbers from 12 to 23 refer to the merged areas in Kyoto. We used values in this figure as the input of the alignment algorithm.

---

[11] https://www.mturk.com/.

	Arashiyama	Fushimi-Inari	Gion-Shijo	Katsura	Kinkaku-ji	Kiyomizu-dera	KyotoImperialPalace	KyotoStation	KyotoUniversity	NijoCastle	NishikiMarket	YasakaShrine
JongwianYai	0.12	0.06	0.1	0.11	0.06	0.06	0.08	0.1	0.07	0.07	0.08	0.09
atchawaram	0.09	0.12	0.05	0.07	0.04	0.18	0.11	0.06	0.08	0.06	0.05	0.09
yMonument	0.08	0.09	0.08	0.09	0.04	0.05	0.1	0.07	0.09	0.08	0.06	0.17
GrandPalace	0.07	0.05	0.05	0.06	0.02	0.05	0.36	0.07	0.07	0.09	0.05	0.07
Silom	0.09	0.05	0.16	0.15	0.05	0.01	0.08	0.09	0.09	0.07	0.11	0.04
Siam	0.08	0.04	0.15	0.08	0.04	0.04	0.13	0.1	0.12	0.03	0.13	0.07
ntainTemple	0.06	0.11	0.05	0.08	0.06	0.14	0.11	0.05	0.08	0.07	0.07	0.11
nUniversity	0.07	0.03	0.06	0.07	0.02	0.03	0.1	0.04	0.41	0.03	0.08	0.05
Chinatown	0.09	0.04	0.07	0.1	0.04	0.03	0.1	0.06	0.09	0.06	0.23	0.1
chakMarket	0.06	0.06	0.06	0.08	0.02	0.02	0.07	0.06	0.07	0.05	0.36	0.08
Asok	0.07	0.03	0.1	0.09	0.03	0.04	0.1	0.25	0.1	0.04	0.09	0.06
Asiatique	0.16	0.08	0.1	0.09	0.07	0.03	0.1	0.06	0.08	0.05	0.11	0.07

Fig. 3. Visualization of answers from Amazon Mechanical Turk. The horizontal scale shows Bangkok places and the vertical scale Kyoto places. The relation between Kyoto places to Bangkok places is presented.

	Asiatique	Asok	ChatuchakMarket	Chinatown	ChulalongkornUniversity	GoldenMountainTemple	Siam	Silom	TheGrandPalace	VictoryMonument	WatArunRatchawaram	WongwianYai
asakaShrine	0.1	0.06	0.01	0.03	0.11	0.15	0.06	0.03	0.08	0.12	0.16	0.09
ishikiMarket	0.1	0.08	0.28	0.08	0.12	0.04	0.1	0.04	0.06	0.04	0.01	0.05
NijoCastle	0.09	0.06	0.02	0.04	0.09	0.11	0.07	0.06	0.22	0.08	0.07	0.09
toUniversity	0.1	0.08	0.01	0.02	0.48	0.06	0.07	0.02	0.06	0.03	0.03	0.05
yotoStation	0.1	0.29	0.03	0.05	0.1	0.06	0.1	0.05	0.05	0.04	0.04	0.08
perialPalace	0.08	0.07	0.01	0.01	0.12	0.1	0.05	0.04	0.31	0.08	0.05	0.07
omizu-dera	0.1	0.1	0.02	0.03	0.09	0.2	0.06	0.04	0.08	0.06	0.18	0.04
Kinkaku-ji	0.11	0.09	0.06	0.05	0.09	0.12	0.08	0.07	0.05	0.07	0.1	0.12
Katsura	0.13	0.07	0.05	0.06	0.1	0.07	0.09	0.1	0.07	0.08	0.06	0.11
Gion-Shijo	0.15	0.12	0.02	0.04	0.1	0.06	0.11	0.07	0.07	0.08	0.06	0.12
ushimi-Inari	0.11	0.09	0.04	0.04	0.07	0.15	0.06	0.05	0.07	0.09	0.14	0.08
Arashiyama	0.17	0.1	0.02	0.03	0.11	0.09	0.07	0.08	0.07	0.07	0.05	0.13

Fig. 4. Visualization of answers from Amazon Mechanical Turk. The horizontal scale shows Kyoto places and the vertical scale shows Bangkok places. The relation between Bangkok places to Kyoto places is presented.

Map Visualization. Map visualizations of Bangkok and Kyoto are shown in Fig. 6. The links of these maps are also available here[12].

Results show that places having similar atmosphere tend to be colored similarly. For illustration, Siam Station and Kyoto Station are the main stations in Bangkok and Kyoto, respectively. They are colored as purple and pinned as "I". The red merged area covers Kyoto University in Kyoto and the Science Center for Education in Bangkok. They might be regarded as similar education places.

---

[12] https://bl.ocks.org/wannita901/raw/6751411e7208f05ce0a2214b9edde79b/.

4.3 Validation

Because data from crowd-sourcing vary according to people's answers, we list them out in ranking, giving 12 similar area rankings for 24 places because they are bi-directional. Then the map from our algorithm and ranking are mutually compared. However, the places in our map are considered by the Google Maps API bounding box. For some places such as Chulalongkorn University, we used the map[13] provided by the university to restrict the bounding box area since the Google Maps API bounding box did not cover the whole area.

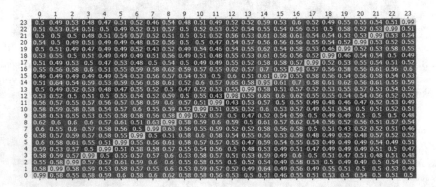

Fig. 5. Correlation matrix of area similarity. In both axes, the area numbers from 0 to 11 represent merged areas in Bangkok, whereas those from 12 to 23 represent merged areas in Kyoto.

The graph in Fig. 7 presents the results of comparing 24 places in the n-rank. The x-axis shows the n-rank step; the y-axis shows the ratio of matched places to all (24 areas). For example, if our map linked Siam with Gion-Shijo as the most similar places and the crowd-sourcing answers voted Siam as the 2nd ranked most similar place to Gion-Shijo, only more than a 2-rank step would be correct. According to the graph, in the 4-rank step, there are 10 among 24 places that match with the answers obtained through crowd-sourcing.

5 Discussion

From our map visualization and the four-rank step validation, we acquired 42% matching with the crowd-sourcing answers, which represented 10 out of 24 pair matches. These correctly mapped areas are found mostly in popular areas with a high number of lively tweets such as Siam Station and Kyoto Station or Silom and Nishiki Market. However, areas with numerous LBSN tweets such as The Grand Palace and Kyoto Imperial Palace were not linked together, which differed from our expectation. The reason might be that these places are likely to have

Fig. 6. Similarity map visualization in Bangkok (left) and Kyoto (right). (Color figure online)

only the checkin tweets as the other sightseeing places, which indicates that the tweet size and variety also affect the results.

Moreover, although we used only 12 candidate places to represent the whole target area, it might not cover all city image and characteristic. For illustration, we expected that Kyoto University and Chulalongkorn University should be linked together, but the result was apparently false. However, we discovered that the area matching Kyoto University is connected with the Science Center for Education, which is also related to education but which is not on the candidate list. However, Chulalongkorn University, which is close to Siam and shopping places, is mapped to Nishiki Market, which is also a popular shopping place. Therefore, the proposed method presents limitations when dealing with places having many functions.

As a result, it might be difficult to say whether similar areas found using the proposed method are correct or not because crowd-sourcing validation might be an overly concrete means of evaluating our map.

Map visualization might be improved in the future if we were to clean the data. For example, to emphasize people's activities in places more, we should use only verbs from tweets to train on Doc2Vec. Improvement of the translator is anticipated as another way of developing our method because the translation process might simplify the text contents.

In addition, because we used the data for only 2 months, we were able to apply these methods with more data or even with data from other cities. Finally,

---

[13] https://www.chula.ac.th/en/contact/map-and-directions/.

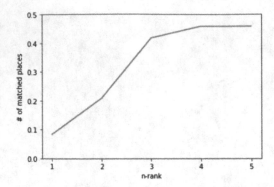

Fig. 7. Map validation with crowd-sourcing.

our experiment was done using Twitter data. Nevertheless, it can apply to use with other geo-tagged texts on social media that provide the exact location.

6   Conclusion

We proposed a method to find similar areas between different cities by exploiting tweets. Specifically we separated tweets in areas into the grid location system and analyzed them using Doc2Vec. Next, we merged similar areas in the respective cities by computing the cosine similarity with various conditions. Then the similar merged areas between the cities were found using the greedy and queue algorithms. The results were visualized on a map. Our results were evaluated by comparison with people's opinions, which were gathered through Amazon Mechanical Turk.

Results demonstrated that the areas in the city center of Bangkok and Kyoto were colored in the same meaning they were likely to be mutually similar, which tends to yield a good outcome. However, the means of validating our map are difficult and must be improved because the responses from crowd-sourcing vary greatly according to people's opinions. Only 12 places might not be sufficient to represent all cities.

Acknowledgements. This work was partly supported by MIC SCOPE #171507010, AMED under Grant Number JP16fk0108119, and JSPS KAKENHI Grant Numbers JP16K16057 and JP16H01722.

References

1. Lynch, K.: The Image of the City. MIT Press, Cambridge (1960)
2. Nold, C.: Greenwich Emotion Map. http://www.emotionmap.net/ (2005)
3. Quercia, D., Schifanella, R., Aiello, L.M.: The shortest path to happiness: recommending beautiful, quiet, and happy routes in the city. In: Proceedings of Conference on Hypertext and Social Media (HyperText) (2014)

4. Quercia, D., Schifanella, R., Aiello, L.M., McLean, K.: Smelly maps: the digital life of urban smellscapes. In: Proceedings of the Ninth International AAAI Conference on Web and Social Media, pp. 327–336 (2015)
5. Aiello, L.M., Schifanella, R., Quercia, D., Aletta, F.: Chatty maps: constructing sound maps of urban areas from social media data. Roy. Soc. Open Sci. **3**(3), 1–19 (2016)
6. Culotta, A.: Towards detecting influenza epidemics by analyzing Twitter messages. In: Proceedings of the First Workshop on Social Media Analytics, pp. 115–122 (2010)
7. Amador Diaz Lopez, J., Collignon-Delmar, S., Benoit, K., et al.: Predicting the Brexit vote by tracking and classifying public opinion using Twitter data. Stat. Polit. Policy **8**(1), 85–104 (2017). Accessed 24 Aug 2018. https://doi.org/10.1515/spp-2017-0006
8. Gerber, M.S.: Predicting crime using Twitter and Kernel density estimation. Decis. Support Syst. **61**, 115–125 (2014). https://doi.org/10.1016/j.dss.2014.02.003
9. Tumasjan, A., Sprenger, T.O., Sandner, P.G., Welpe, I.M.: Predicting elections with Twitter: what 140 characters reveal about political sentiment. In: International AAAI Conference on Weblogs and Social Media, Washington, DC (2010)
10. Hahmann, S., Purves, R., Burghardt, D.: Twitter location (sometimes) matters: exploring the relationship between georeferenced tweet content and nearby feature classes. J. Spat. Inf. Sci. Number **9**, 1–36 (2014)
11. Preotiuc-Pietro, D., Cranshaw, J., Yano, T.: Exploring venue-based city-to-city similarity measures. In: Proceedings of the Second ACM SIGKDD International Workshop on Urban Computing, Article No. 16 (2013)
12. Kato, M.P., Hiroaki, O., Oyama, S., Tanaka, K.: Search as if you were in your home town: geographic search by regional context and dynamic feature-space selection. In: Proceedings of the 19th ACM International Conference on Information and Knowledge Management, pp. 1541–1544 (2010)
13. Seth, R., Covell, M., Ravichandran, D., Sivakumar, D., Baluja, S.: A tale of two (similar) cities: inferring city similarity through geo-spatial query log analysis. In: Proceedings of the International Conference on Knowledge Discovery and Information Retrieval (2011)

A Visual Analytics Approach to Exploration of Hotels in Overlaid Drive-Time Polygons of Attractions

Chong Zhang[1,2]($\boxtimes$), Zhengcong Yin[3], Peng Gao[2], and Sathya Prasad[2]

[1] University of North Carolina at Charlotte, Charlotte, NC 28262, USA
[2] Environmental Systems Research Institute, Inc., Redlands, CA 92373, USA
czhang@esri.com
[3] Texas A&M University, College Station, TX 77843, USA

Abstract. Given multiple attractions to go to, choosing hotels is a Bichromatic Reverser k Nearest Neighbor searching problem. Many efforts have been made to address the problem in the domain of spatial databases. However, most of them suffer from the difficulty of visual selection and comparison. To integrate human intuition into the analysis process and facilitate information communication with users, geospatial visualization techniques have been developed. Unfortunately, too much information on the map, such as hotels and attractions, and drive-time polygons, could overwhelm users. Following the focus+context principle of information visualization, we propose a visual analytics approach to provide users with the flexibility of making their hotel choice. It consists of three main components: an overlap-free and space-efficient clustering algorithm to reduce the searching space for attractions in the map view (MapView), visual comparison of attraction-hotel drive-time in brushable Small Multiples View (SMView), and the coordination between the two views. We demonstrate the utility of this approach and explain how MapView and SMView can help address the problem.

Keywords: Reverse nearest neighbor · Visual analytics · Spatial clustering · Overlap-free

1 Introduction

While choosing a hotel is a multi-criteria decision (cost, rating, class, facility, transportation, etc.) at a conceptual level, the problem based on spatial proximity is still interesting. The hotels selected based on spatial proximity are often considered as a superset of the final one that user will choose to stay. It's not an easy task. If the preference of these attractions are equally considered, the ideal hotel location candidates should be near the "center" of the polygon connecting each attraction such that they have the same and shortest distance to the

---

Live demo of the interface is available here.
This work is part of Chong Zhang's internship at ESRI.

© Springer Nature Switzerland AG 2019
Y. Kawai et al. (Eds.): W2GIS 2019, LNCS 11474, pp. 28–40, 2019.
https://doi.org/10.1007/978-3-030-17246-6_3

attractions. The "center", however, might not exist or not be easily found: 1. it is a NP-hard geometric optimization problem [18], 2. the polygon connecting all attractions could be complex, irregular, or self-intersecting which would result in no such center, and 3. too much information on a map overwhelms users. Therefore, there are no universal methods to calculate the "center" for all possible travel paths.

Reverse k-Nearest Neighbor (RkNN) query has been utilized to address the complementary problem. It has been formally defined by [15]. For a query point of type A, the objective of RkNN is to find data points of type B in which the query point is their nearest A point. When the types of A and B are different, the query becomes Bichromatic RkNN (BRkNN for short). For example, it can be used to find the top k nearest hotels to stay when you have certain attractions to go to.

In practice, from the user's perspective, choosing a hotel is a progressive process where the user can switch back and forth between the selections of attractions and adjust his interest based on the hotel candidates provided by the system. In the domain of spatial databases, while many efforts have been made to develop RkNN searching algorithms [14,21,26,28], most of them suffer from the difficulty of visual selections and comparisons.

In the domain of geospatial visualization, many techniques have been developed to deal with overcrowded displays for a better communication with users [9]. Since attractions are very close to each other for some area, any of these attractions can be a starting point in the drive-time analysis (see Sect. 3) for the travel time estimates. Therefore, spatial clustering is commonly used to aggregate a set of attractions. Common spatial point clustering approaches such as grid-based [27], Rtree based [11], Quadtree based [2] often present over-clustering or overlapping result (see Fig. 1) when the size of visual element (like circle) encodes the number of points in cluster (cluster size). Aggregating data points too much might result in information loss and misleading hypothesis. On one hand, users expect to see more rich information on an empty space. If there are no clusters displayed for a particular area, users might believe that the data for that area is not available. On the other hand, early merging neighbors might lead to an incomplete understanding of distribution patterns over the area.

Styling doesn't help much with information overload. If too much information on the map, such as drive-time polygons, attractions, and hotels, is styled using different colors or symbols, the resulting visual clutter could prevent users from understanding the map. Furthermore, rendering a web map with a large number of points on a single view is a challenging work to a processor and can be unresponsive. Therefore, existing works focus on multiple views [13,25]. While these views enable users to emphasize different information, they often suffer from the separation of focus and context [16,22]. Many users can be lost in the transition of information from one view to another.

In this paper, we propose a visual analytics approach to exploration of hotels that fall within overlaid drive-time polygons. The main contribution of the paper is as follows:

- We propose a novel space-efficient and overlap-free spatial point clustering algorithm. This algorithm allows geometrical representation of clusters with varied radii.
- We design a brushable SMView of scatterplot for hotels that fall within drive-time polygons. Brushing on one scatterplot selects, highlights, and colors corresponding hotels in other scatterplots in the SMView.
- We present an interactive and focus+context visualization tool with two coordinated views to deal with BRkNN searching problems. MapView displays the locations of attractions and hotels as well as the drive-time polygons. SMView preserves geographical placements of the clicked attraction and hotels in the drive-time polygon. Two views are linked and have the consistent color coding.

To the author's knowledge, this study is the first attempt to leverage visual analytics to address BRkNN searching problems.

2   Related Work

Many earlier efforts have been made in the domain of spatial database to develop RkNN searching algorithms. For example, [21] divide the space into six regions using three intersecting lines at an angle of 60°. The candidates then are on a circle with the intersected center. However, the number of returns is limited to six. While the Incremental and General Evaluation of continuous Reverse Nearest neighbor queries, an algorithm proposed in [14], can handle BRkNN case and overcome the limitation, it aims at the processing of continuous RkNN queries on moving objects. Recent techniques commonly utilize a pruning-and-verification framework for RkNN and BRkNN queries [26,28].

In practice, from the user's perspective, BRkNN query could be a progressive process where the user can switch back and forth between the point of interest (POI) selections of type A and type B or even make a compromise between them. Most of RkNN searching algorithms have not considered leveraging of interactive and coordinated visualizations which can aid to make the selections and comparisons.

GeoSpatial visualization techniques help make visual comparisons and decisions [5,23]. When a map is used to convey rich information, coordinated views are commonly utilized [12,20]. Scatterplot [6] is one bivariate exploration method. Small Multiples [24] enable comparison across variables and reveal potential patterns that are difficult to find on the map.

Spatial point clustering is also commonly used to mitigate the overcrowded problem [9]. Common spatial point clustering approaches include partitioning (k-means [10], k-medoids [19]), hierarchical [17], density-based (DBSCAN [1]), grid-based [27], Rtree based [11], and Quadtree based [2]. Although they have different application scenarios, all share a problem of cluster overlapping when size of visual element (like circle) encodes the number of points in cluster (cluster size).

3   Drive-Time Analysis

Drive-Time Analysis models the driving time from a location. It utilized digitized street networks, speed limits, travel modes (car, walk, etc.), traffic conditions, and date and time. These parameters describe the impedance of traveling along the road segments. It is an important analysis in planning tasks. For example, it can be used to determine how much of the city is within a three-minute drive of the city's fire stations in heavy and light traffic.

Given a starting point location, Dijkstra's algorithm [8] is typically used to traverse the street network and return edges that are within the specified travel time. Travel polygons are then generated by connecting the edges into a Triangulated Irregular Network data structure. The polygons represent the extent to which a vehicle/person can travel outward from the location in all directions.

4   Approach Overview

We develop two coordinated views: MapView and SMView. MapView dynamically displays the locations of attractions and hotels, drive-time polygons, and visual hints including labels for clicked attractions and highlightings for hotel candidates. We propose a space-efficient and overlap-free point clustering algorithm. This algorithm is integrated into the attraction point layer (a kind of D3Layer). Dynamic clustering process is provided so that users can zoom to see changes in the level of details. On the SMView, each scatterplot represents the hotels that fall within the corresponding drive-time polygon. The geographic placements of these hotels are remained on the SMView. Moreover, we employ the brushing interaction to each scatterplot and allow users to examine the drive-time differences across other clicked attractions. Two views are coordinated with the same color coding: green means less than 1 min, yellow means less than 2 min, and dark blue is within 3 min.

5   Overlap-Free and Space-Efficient Point Clustering

Common grid-based point clustering algorithms used in Google Maps[1], Mapbox[2], or Leaflet[3] presents a great performance. However, it either suffers from space inefficiency (over clustering) or overlapping (see Fig. 1), especially when the radius of a circle is used to represent the cluster size (the more attractions in the cluster the larger the circle will become). The data points of attractions in Fig. 1(a) could have been more aggregated. While the clusters in Fig. 1(b) could have expanded a little to utilize spaces presenting more information.

---

[1] https://googlemaps.github.io/js-marker-clusterer/.
[2] https://www.mapbox.com/mapbox-gl-js/example/cluster/.
[3] https://github.com/Leaflet/Leaflet.markercluster.

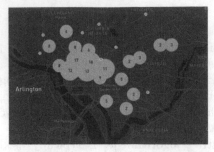

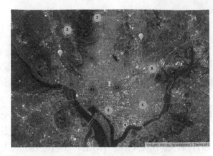

(a) Overlapping occurs using Mapbox GL JS API

(b) Over-clustering/space-inefficient occurs using Google Maps API

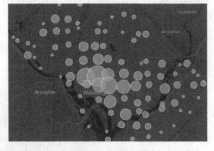

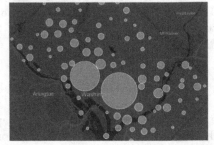

(c) Space-efficient but still overlapping using Quadtree based clustering

(d) Space-efficient and overlap-free clustering using the proposed algorithm

Fig. 1. Comparison between common clustering algorithms with the same dataset.

We propose a new space-efficient and overlap-free point clustering algorithm. The algorithm consists of three components: a grid system of the attractions for spatial efficiency tolerance, a Quadtree of the attractions for finding the initial cluster centers, and a Delaunay Triangulation (DT) [7] graph of attractions, or clusters of attractions for proper merging (see Fig. 2). The two components are applied as the map extent gets changed (e.g. zoom level).

The details of the algorithm is described as follows. First, we define a grid system for the whole map. The size of the grid cell is related to the space efficiency tolerance (called SECell, see line 2 in Algorithm 1). It is used for Quadtree **lookup** (see line 14 in Algorithm 1). This number determines the degree of details revealing for each map zoom level. The smaller the number is, the more points are drawn. Initially, we assign each attraction to its own cluster (see line 3 in Algorithm 1). Then, for all attractions found in a SECell, we **merge** them to a new cluster (see line 16 in Algorithm 1) with a weighted radius of logarithmic scale or square root scale depending on the degree of separation of clusters (see line 10 in Algorithm 1). This step continues until we traverse all SECells.

For the new cluster within each SECell, if radius of circle is used to represent the cluster size, overlapping might occur for some neighboring SECells (see Fig. 1(c)). Therefore, the second step is to **merge** the overlapped clusters (see line 23 in Algorithm 1). Instead of brute-force searching the closer clusters, we

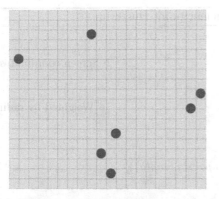

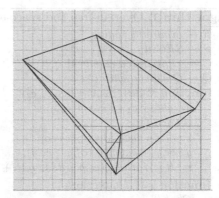

(a) Original attractions within a grid system.

(b) Create a Quadtree and a DT graph for these attractions.

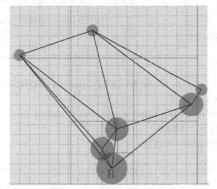

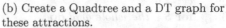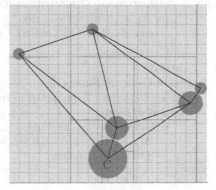

(c) Render these attractions with different sized circles. Overlapping between A and B is detected.

(d) Only merge A and B into C for this clustering process.

Fig. 2. The main step of the proposed algorithm

take advantage of the structure of DT and check which clusters are overlapping that should be merged into a new cluster (see line 22 in Algorithm 1). In a sense, overlapping is caused by the distance of two clusters being less than the sum of their radii. If we generate a DT graph using the center coordinates of these clusters, we just need to compare the length of one edge and the sum of the radii of the connecting clusters (see Fig. 2(b)).

6   Small Multiples of Brushing-Enabled Scatterplots

Each attraction creates a drive-time polygon. This polygon has three concentric regions indicating the driving time of 1, 2, or 3 min. These three regions are then color-coded in green, yellow, and blue, respectively (see MapView of Fig. 4). Clicking multiple attractions creates multiple drive-time polygons which might

Algorithm 1. Overlap-free and space-efficient point clustering

1: $R \leftarrow mininum\ radius$
2: $S \leftarrow SECell$
3: $C \leftarrow \{initialized\ clusters\}$ ▷ Screen coordinates
4: $Q \leftarrow Quadtree\ with\ C$
5: **function** Merge_Clusters(c_1, c_2)
6:     $weight \leftarrow c_2.r/(c_1.r + c_2.r)$ ▷ Weighted by radii
7:     **return** {
8:     $x : c_1.x + weight * (c_2.x - c_1.x)$,
9:     $y : c_1.y + weight * (c_2.y - c_1.y)$,
10:    $r : R * Math.log(c_1.points.length + c_2.points.length)\}$
11: **function** Init-Clustering()
12:    $C' \leftarrow []$
13:    **for** $S \in the\ whole\ grid$ **do**
14:        $attractions \leftarrow Q.search(S.x, S.y)$
15:        **for** $a_1, a_2 \in attractions$ **do**   ▷ Reduce operation
16:            $ave \leftarrow MERGE\_CLUSTERS(a_1, a_2)$
17:        $C'.push(ave)$
18:    **return** C'
19: **function** OverlapFree(C')   ▷ C' from INIT-CLUSTERING()
20:    $DT \leftarrow Delaunay\ Triangulation\ links\ with\ C'$
21:    **for** $L \in DT$ **do**   ▷ L consists of a source node and a target node.
22:        **if** $L.distance < L.source.r + L.target.r$ **then**
23:            $MERGE\_CLUSTERS(L.source, L.target)$
24:            $Delete\ C[L.target]$

overlap each other with the same color codings (see MapView of Fig. 5). This leads to reading problems: 1. hotels that fall within these polygons are very difficult to find; 2. color coding for these hotels could be impossible when they fall in different time zones in the overlaid drive-time polygons. For example, in the MapView of Fig. 5, multiple different drive-time polygons are overlapped. Some hotels falling within the 1-min time zone in one polygon also fall within the 3-min time zone in another polygon.

To overcome these problems, we need to bring users' analysis space to a separate view while keeping the context. We design a Small Multiples view of scatterplot for hotels within the drive-time polygon (see Fig. 3). A single scatterplot is a simplified clone of the clicked attraction and corresponding hotels that fall within the drive-time polygon. The x axis and y axis of scatterplot represent projected coordinates of hotels. Therefore, the geographic references of these hotels are remained on the view. The color used in the scatterplot is consistent with drive-time polygons: green means 1 min, yellow means 2 min, and blue is 3 min. Moreover, the clicked attraction is presented in this view as the same color (blue) in the MapView (see the SMView of Fig. 4).

Once users click multiple attractions, corresponding scatterplots are generated in the SMView (see the SMView of Fig. 5). We employ brushing to each

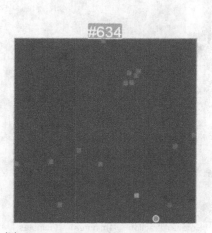

(a) Drive-time polygon for attraction #634.

(b) Scatterplot for the clicked attraction #634.

Fig. 3. Hotels falling within the drive-time polygon are projected to a scatterplot in the SMView with geographic placement remained. With the same color coding for the drive-time differences, the clicked attraction is labeled on the top of the scatterplot and its location is also highlighted in the same legend as MapView. (Color figure online)

scatterplot which allows users to compare the driving times across other clicked attractions. For example, in the SMView of Fig. 6, by using the brush on the first scatterplot (attraction #304) to select hotels in the box, these hotels get color-coded based on their drive-times at the clicked attraction, and all other scatterplots also clearly highlight the relative locations of these hotels and color-codes the drive-times at corresponding clicked attraction. By reading the color coding, brushed hotels around attraction #304 are less than 1 min away. However, these hotels are more than 2 min away from #49.

7 Implementation

Drive-Time Analysis. In this paper, we use Esri's Network Analysis Services[4] to create the drive-time polygons on the fly. The input layer is attraction point features. The resulting drive time polygons are visualized based on the driving minutes (1, 2, or 3 min). Hotels that fall within these polygons can be queried using a geoprocessing operation of spatial intersects.

D3Layer. D3 [3] is used to render GeoJSON datasets and provide a better interaction between two views. D3Layer is developed on top of ArcGIS API for

---

[4] https://developers.arcgis.com/rest/network/api-reference/service-area-asynchronous-service.htm.

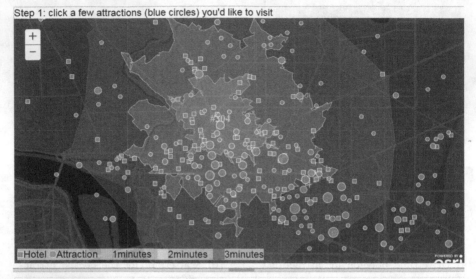

Fig. 4. Selecting one attraction *#304* in the MapView creates one scatterplot for it in the SMView. MapView has only attraction data clustered. Circle radius represents the cluster size. (Color figure online)

JavaScript for the clustering algorithm implementation and GeoJSON dataset rendering. It extends *"esri/layers/GraphicsLayer"* in which the life-cycle methods are invoked. We fully take advantage of the rendering pipeline in ArcGIS API for JavaScript to provide a great user experience for the system. The layer offers a dynamic clustering. For every time map gets updated such as panning, zooming, the layer starts the clustering algorithm and renders the cluster results on the browser end.

Data Source. Attraction data source is from ArcGIS Hub[5]. Hotel data source is from ArcGIS Hub[6]. Both of them are in the format of GeoJSON [4].

---

[5] http://hub.arcgis.com/datasets/9927e456ac024b11811323812934edbb_12.
[6] http://hub.arcgis.com/datasets/a3ed163dbf994792a010d742ef1f683d_6.

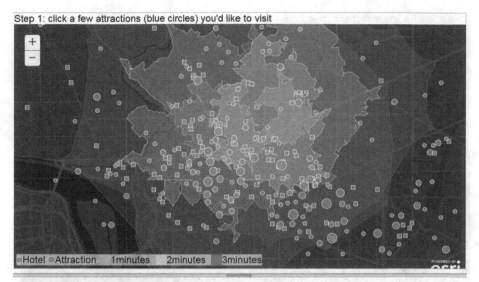

Step 2: click-and-drag hotels (squares) to compare driving times with the same color coding.

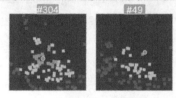

Fig. 5. Two attractions selected and two corresponding scatterplots created. (Color figure online)

8 Interaction Design

An important role of the approach is to enable an interactive exploration and validation of interesting hotels. In our approach, these processes are facilitated by a set of interactions on both MapView and SMView. In particular, a set of interactions are provided on the MapView. First, mouse-hover interaction displays the details of hotels in tooltips (see the MapView of Fig. 6). Users can also hover over a blue circle (cluster of attractions) to trigger a tooltip display of the detailed information, such as the cluster size, the attraction Identity Number (ID), and the attraction names (only the first three entities). Users can click any blue circle to create the drive-time polygons. Since the clicked blue circle might represent more than one attractions, the closest attraction to the mouse position is chosen. The ID of the clicked attraction is shown on the mouse position with a prefix "#". Meanwhile, the clicked attraction is highlighted using green square underneath it (see the symbol in the MapView of Fig. 4). In order to support the legibility for the polygons, different outline colors are used from inside fill. Only hotels that fall within the drive-time polygon are selected into the SMView. User

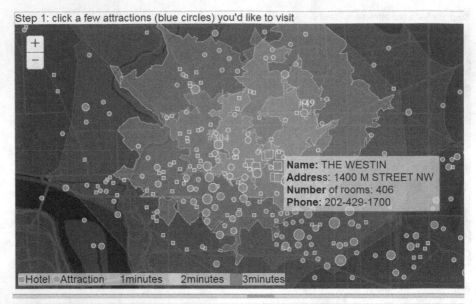

Step 1: click a few attractions (blue circles) you'd like to visit

Name: THE WESTIN
Address: 1400 M STREET NW
Number of rooms: 406
Phone: 202-429-1700

Hotel Attraction   1minutes   2minutes   3minutes

Step 2: click-and-drag hotels (squares) to compare driving times with the same color coding.

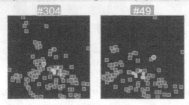

Fig. 6. Brushing on SMView with two scatterplots. MapView shows that two attractions are clicked and two corresponding drive-time polygons are created. Hotels selected in the brushing box in the SMView are highlighted as orange squares with a resize animation. SMView shows that brushing is on the scatterplot of attraction *#304*. The other scatterplot for attraction *#49* applies the brushing selection and update the color coding for the brushed hotels if identified. (Color figure online)

can click multiple blue circles, create multiple drive-time polygons, and conduct the overlay analysis, as shown in Fig. 5.

In the SMView, users can move the brushing box to change the hotel selections. As shown in Fig. 6, only hotels within the box are color-coded. Hotels located outside the box are greyed out to reduce the distraction. Coordinately, the MapView highlights these selected hotels as orange squares with an animation of size changes, as shown in Fig. 6. The size of the brushing box can be changed. With the hint of the moving cursor, users can drag the corner of the brushing box to expend or reduce the selection range. Moreover, the brushing box can be moved to change the selection. While it is moving, the color coding of the selections gets real-time updated. Coordinately, the MapView also updates the animation of the selected hotels as well as the viewing center. Clicking outside the brushing box cancels the selection.

9 Conclusions and Future Work

We present a visual analytics approach to exploration of hotels that fall within overlaid drive-time polygons. We propose an overlap-free and spatial-efficient point clustering algorithm, employ brushing to SMView that interact with MapView, and demonstrate a visual analytics system with two coordinated views. This system provides travelers with an interactive exploration of hotels overlaid in drive-time polygons of attractions. The proposed approach can also be applied to other types of BRkNN problems.

Although we have gathered a software development director, two software development managers, a few PhD students, and engineers for reviewing, and have received encouraging feedbacks in a mid-size software company, it will be critical to conduct a serious evaluation aimed at quantifying the performance of the algorithm and the effectiveness of the collaborative interface.

Because of Multi-Criteria Spatial Decision Support (MCSDS) system that is being employed for site selection where several factors can be weighted or combined, we would also like to incorporate our system with MCSDS system to provide a better interactive visual multi-criteria spatial decision analysis.

References

1. Birant, D., Kut, A.: ST-DBSCAN: an algorithm for clustering spatial-temporal data. Data Knowl. Eng. **60**(1), 208–221 (2007)
2. Bishnu, P.S., Bhattacherjee, V.: Software fault prediction using quad tree-based k-means clustering algorithm. IEEE Trans. Knowl. Data Eng. **24**(6), 1146–1150 (2012)
3. Bostock, M., Ogievetsky, V., Heer, J.: D^3 data-driven documents. IEEE Trans. Visual. Comput. Graph. **17**(12), 2301–2309 (2011)
4. Butler, H., Daly, M., Doyle, A., Gillies, S., Schaub, T., Schmidt, C.: The GeoJSON format specification. Rapport technique **67** (2008)
5. Cartwright, W., et al.: Geospatial information visualization user interface issues. Cartogr. Geogr. Inf. Sci. **28**(1), 45–60 (2001)
6. Cleveland, W.C., McGill, M.E.: Dynamic Graphics for Statistics. CRC Press, Inc., Boca Raton (1988)
7. Lee, D.-T., Schachter, B.J.: Two algorithms for constructing a Delaunay triangulation. Int. J. Comput. Inform. Sci. **9**(3), 219–242 (1980)
8. Dijkstra, E.W.: A note on two problems in connexion with graphs. Numerische mathematik **1**(1), 269–271 (1959)
9. Ellis, G., Dix, A.: A taxonomy of clutter reduction for information visualisation. IEEE Trans. Visual. Comput. Graph. **13**(6), 1216–1223 (2007)
10. Fahim, A., Salem, A., Torkey, F.A., Ramadan, M.: An efficient enhanced k-means clustering algorithm. J. Zhejiang Univ.-Sci. A **7**(10), 1626–1633 (2006)
11. Han, J., Miller, H.J.: Geographic Data Mining and Knowledge Discovery. CRC Press, Boca Raton (2009)

12. Huang, X., Zhao, Y., Ma, C., Yang, J., Ye, X., Zhang, C.: TrajGraph: a graph-based visual analytics approach to studying urban network centralities using taxi trajectory data. IEEE Trans. Visual. Comput. Graph. **22**(1), 160–169 (2016)
13. Hubmann-Haidvogel, A., Scharl, A., Weichselbraun, A.: Multiple coordinated views for searching and navigating web content repositories. Inf. Sci. **179**(12), 1813–1821 (2009)
14. Kang, J.M., Mokbel, M.F., Shekhar, S., Xia, T., Zhang, D.: Continuous evaluation of monochromatic and bichromatic reverse nearest neighbors. In: IEEE 23rd International Conference on Data Engineering, ICDE 2007, pp. 806–815. IEEE (2007)
15. Korn, F., Muthukrishnan, S.: Influence sets based on reverse nearest neighbor queries. In: ACM SIGMOD Record, vol. 29, pp. 201–212. ACM (2000)
16. Mindek, P., Gröller, M.E., Bruckner, S.: Managing spatial selections with contextual snapshots. In: Computer Graphics Forum, vol. 33, pp. 132–144. Wiley (2014)
17. Ng, R.T., Han, J.: CLARANS: a method for clustering objects for spatial data mining. IEEE Trans. Knowl. Data Eng. **14**(5), 1003–1016 (2002)
18. Owen, S.H., Daskin, M.S.: Strategic facility location: a review. Eur. J. Oper. Res. **111**(3), 423–447 (1998)
19. Park, H.S., Jun, C.H.: A simple and fast algorithm for k-medoids clustering. Expert Syst. Appl. **36**(2), 3336–3341 (2009)
20. Roberts, J.C.: State of the art: coordinated & multiple views in exploratory visualization. In: Fifth International Conference on Coordinated and Multiple Views in Exploratory Visualization, CMV 2007, pp. 61–71. IEEE (2007)
21. Stanoi, I., Agrawal, D., El Abbadi, A.: Reverse nearest neighbor queries for dynamic databases. In: ACM SIGMOD Workshop on Research Issues in Data Mining and Knowledge Discovery, pp. 44–53 (2000)
22. Stasko, J., Zhang, E.: Focus+ context display and navigation techniques for enhancing radial, space-filling hierarchy visualizations. In: IEEE Symposium on Information Visualization, InfoVis 2000, pp. 57–65. IEEE (2000)
23. Takatsuka, M., Gahegan, M.: GeoVISTA Studio: a codeless visual programming environment for geoscientific data analysis and visualization. Comput. Geosci. **28**(10), 1131–1144 (2002)
24. Tufte, E.R.: Envisioning Information. Graphics Press, London (1990)
25. Walsh, J.A., Zucco, J., Smith, R.T., Thomas, B.H.: Temporal-geospatial cooperative visual analysis. In: Big Data Visual Analytics (BDVA), pp. 1–8. IEEE (2016)
26. Wang, S., Cheema, M.A., Lin, X., Zhang, Y., Liu, D.: Efficiently computing reverse k furthest neighbors. In: 2016 IEEE 32nd International Conference on Data Engineering (ICDE), pp. 1110–1121. IEEE (2016)
27. Wang, W., Yang, J., Muntz, R., et al.: STING: a statistical information grid approach to spatial data mining. VLDB **97**, 186–195 (1997)
28. Yang, S., Cheema, M.A., Lin, X., Zhang, Y., Zhang, W.: Reverse k nearest neighbors queries and spatial reverse top-k queries. VLDB J. **26**(2), 151–176 (2017)

Identifying Preferred Areas
in Road Networks

Stefan Funke[(✉)]

Universität Stuttgart, Stuttgart, Germany
`funke@fmi.uni-stuttgart.de`

Abstract. We consider the problem of identifying preferred areas on a map which are characterized by a conjunction of minimum or maximum distance conditions from or to (classes of) points of interest. By applying state-of-the-art speed-up techniques for shortest path queries we achieve interactive response times allowing for an efficient web application even for country sized maps.

Keywords: Points of interest · Shortest path distances

1 Introduction

Consider the following problem in a map context: Assume you wish for a career change and have already acquired several job offers in different cities within Germany. But typically the decision for one of these job offers does not only depend on the job itself but also on possible places of residence near the workplace and criteria like proximity of shopping locations, children daycare, or sports/health facilities. Apart from these rather 'local' criteria, you might also prefer places of residence which are not too far from your parents' home but at the same time not too close to your parents-in-law's home. More formally, if L_w, L_s, L_d, L_f are the sets of locations of the workplaces, supermarkets, children daycare facilities, and health/fitness clubs, respectively, we might be interested in the locations in Germany which have *simultaneously* at least one element of L_w within 30 min, one element of L_s within 15 min, one element of L_d within 10 min, and one element of L_f within 5 min driving time, but additionally are at most a 2 h drive from your parents' home but at least a 4 h drive from your parents-in-law's home away.

In this paper we show how to efficiently compute such locations even on a country-sized road network and present the results as an interactive webservice.

2 Preliminaries and Related Work

Our proposed solution approach relies on the *contraction hierarchy* (CH) [2] data structure and the PHAST scheme, both developed to speed up shortest path queries. The CH augments a given graph with *shortcuts* and a function *level* :

© Springer Nature Switzerland AG 2019
Y. Kawai et al. (Eds.): W2GIS 2019, LNCS 11474, pp. 41–45, 2019.
https://doi.org/10.1007/978-3-030-17246-6_4

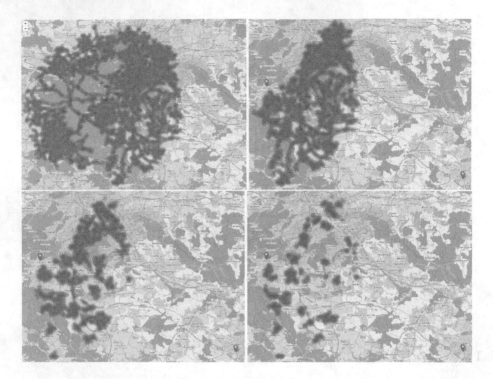

Fig. 1. Preferred areas: ≤ 15 min from University Stuttgart (top left), additionally ≤ 30 min from the town of Kirchheim (top right), additionally ≤ 3 min from a supermarket (bottom left), and additionally ≤ 2 min from a school.

$V \to \mathbb{N}$. Based on that, source-target queries can be performed about 3 orders of magnitude (milliseconds vs seconds) faster than plain Dijkstra by restricting the search to unimodal (wrt node levels) paths between source and target. The PHAST scheme [1] builds upon a constructed CH and allows for efficient one-to-all distance computations, as is briefly reviewed in the next section.

3   Preferred Locations

As described in the introduction we aim at identifying all nodes in the network that simultaneously satisfy our minimum or maximum distance conditions. Basically we will treat each condition separately, computing for every node a distance value for each condition, and then with a final sweep over all nodes of the network determine all 'preferred locations' satisfying all conditions. Let us first consider conditions which involve only one point of interest like the maximum distance from your parents' home. Here the naive approach is to simply perform one single Dijkstra run to compute distances to all nodes of the network. Note that for our road network of Germany this takes up to around 10 s on a single core of a modern CPU without much potential for speedup via parallelization.

Fortunately, the PHAST approach is tailor-made exactly for that purpose. It allows to compute one-to-all distances within less than a second by in a first phase performing a Dijkstra search from the source but only considering edges leading to higher-level nodes. Then in a second phase all edges are considered and relaxed in decreasing order of the target level. The speedup of around a factor of 10 compared to standard Dijkstra is mainly due to the well-behaved memory access pattern. If one-to-all distances from more than one source are desired, one still requires only one single sweep over the edges. The running time for computing distances from two sources is then hardly more than that for one source only.

Distance to Sets of POIs – Naive and Accelerated Computation. If the condition involves a *set* of points of interest instead of just a single POI, a naive approach would be to simply run one Dijkstra from each POI and then for each node in the network take the minimum over all computed distances. With more than 30,000 supermarkets in Germany this is obviously infeasible, even with a PHAST-accelerated approach.

For the plain Dijkstra's algorithm, a possible improvement is to insert *all* POIs (e.g. all nodes associated with a supermarket) simultaneously into the priority queue with a distance of 0 right at the beginning. Then one single Dijkstra run suffices to compute for each node in the network the distance to its nearest supermarket. While being a drastic improvement compared to the naive approach, it is still too slow for application in a webservice. But we can modify the PHAST scheme to the same effect and avoid unnecessary distance computations. We also simply insert all POIs with distance 0 into the priority queue in the first phase of the PHAST scheme. Note that the second phase – the linear scan over and relaxation of all edges – is not affected at all. Since the latter also dominates the running time of the PHAST scheme, the total running time hardly increases as we will see in our experimental evaluation, unless we have a really large set of POIs.

4   Implementation and Evaluation

Let us first evaluate the basic building blocks of our system, namely the efficient computation of distances from (a set of) points of interests. Our implementation was written in C++, compiled using g++ 7.3.0, and evaluated on an AMD Ryzen 1800x (a sub 200$ CPU) running Ubuntu 18.04.

Benchmarking. We extracted the road network of Germany from the OpenStreetMap project with travel times as edge costs. This resulted in a graph with 26,857,972 nodes and 54,227,923 edges; a CH constructed afterwards increased the latter to 97,771,039 and took less than 10 min.

In the left part of Table 1 we have analyzed the cost of simple one-to-all distance computations (as necessary for the conditions involving your parents' and your parents-in-law's homes). As mentioned before, in PHAST one can combine two or more one-to-all computations such that only one sweep over the edges is necessary, hence the running time for two sources in fact is hardly more than that for a single source. For two sources, the speedup compared to plain Dijkstra is about a factor of 20.

Table 1. Benchmark results (in seconds, averaged over 10 random instances).

	One-to-all		Set distance				
	1 src	2 srcs	$|P| = 10$	$|P| = 100$	$|P| = 1000$	$|P| = 10,000$	$|P| = 100,000$
Dijk	10.0	19.1	12.2	15.6	17.4	18.4	19.9
PHAST	0.83	0.91	1.19	1.19	1.21	1.33	1.97

In the right part of Table 1 we have analyzed the cost of setDistances for varying set sizes. We expect smaller set sizes in conditions, e.g., on the distance to possible workplaces, whereas conditions on the distance to supermarkets or daycare facilities typically involve set sizes in the order of thousands. For benchmarking, the respective sets were always chosen uniformly at random from the whole network. It turns out that running times are hardly affected by the set size (since the dominating sweep over all edges is unaffected by the set size) – ideal for our concrete application scenario. Only for very large set sizes, the first phase of the modified PHAST becomes noticeable.

Demonstrator. Our application consists of a server component, which contains the C++ backend where the above distance computations are performed together with a network interface based on boost.beast. The leaflet-based web page interacts via web sockets with the server component, allows for specification of maximum and minimum distance conditions, and displays the result as a leaflet heatmap. See Fig. 1 as an example where we show preferred areas when adding more and more distance conditions. The computation time for this type of queries was always below two seconds independent of the location of POIs. Data about points-of-interest like supermarkets, schools, etc. were extracted from OpenStreetMap.

5   Conclusions and Outlook

We have shown how state-of-the-art speed-up techniques for shortest path queries allow for quickly determining 'preferred areas' in a road network. Of course, typically there are more criteria than just distances to points-of-interest that make an area preferable, e.g., housing demographics or scenicness. It remains to extend our basic approach to also take into account such criteria.

References

1. Delling, D., Goldberg, A.V., Nowatzyk, A., Werneck, R.F.: PHAST: hardware-accelerated shortest path trees. J. Parallel Distrib. Comput. **73**(7), 940–952 (2013)
2. Geisberger, R., Sanders, P., Schultes, D., Vetter, C.: Exact routing in large road networks using contraction hierarchies. Transp. Sci. **46**(3), 388–404 (2012)

How to Extend IndoorGML for Seamless Navigation Between Indoor and Outdoor Space

Taehoon Kim[1], Kyoung-Sook Kim[1(✉)], and Jiyeong Lee[2]

[1] National Institute of Advanced Industrial Science and Technology (AIST),
Tokyo, Japan
{kim.taehoon,ks.kim}@aist.go.jp
[2] The University of Seoul, Seoul, Republic of Korea
jlee@uos.ac.kr

Abstract. Navigation and tracking systems became a traditional tool of positioning determination and route guidance for moving objects such as vehicles and pedestrians. Position techniques like Global Positioning Systems (GPS) and map information about navigation space are fundamental elements in these systems. As matching a position into the map, a route plan is determined by navigational knowledge of space constraints. In this paper, we propose an extension model of Open Geospatial Consortium (OGC) IndoorGML for representing a seamless navigation space, called Seamless Navigation Model (SNM). In particular, we link the existing spatial models for seamless navigation between indoor and outdoor environments for not only humans but also robots instead of developing a unified model of indoor and outdoor spaces. This study presents how to realize the concept of Anchor node in IndoorGML and define characteristics to connect different geospatial models of indoor and outdoor spaces. Finally, we provide a use-case of SNM with a road network represented by OGC CityGML, which is a standard format for the storage and exchange of virtual three-dimensional city models. Our model can reduce the integration cost of different geospatial models and encourage the reuse of the current map information for various tasks not only seamless navigation.

Keywords: Anchor node · Seamless navigation space ·
OGC IndoorGML · OGC CityGML

1 Introduction

With the advances in positioning and wireless communication technologies, navigation systems have become one of the most widely used types of location-based services in not only outdoor but also indoor space. An accurate positioning and map information are essential for the route guidance of objects moving from one

© Springer Nature Switzerland AG 2019
Y. Kawai et al. (Eds.): W2GIS 2019, LNCS 11474, pp. 46–62, 2019.
https://doi.org/10.1007/978-3-030-17246-6_5

place to another. As matching a position into the map, a route plan is determined by navigational knowledge of space constraints. Comparing to a long history of outdoor navigation, indoor navigation has recently witnessed an increase in interest with new positioning systems (e.g., WiFi, Bluetooth, and inertial navigation system) and indoor maps (e.g., Google Indoor[1], OpenStreetMap (OSM) Indoor[2], and Apple IMDF[3]). However, these indoor maps are based on two-dimensional floor plans for the position mapping like outdoor maps and bring issues of interoperability because of their different spatial models.

There are three major standards that provide spatial data models and formats for the representation and exchange of indoor space: IFC [3] of buildingSMART, OGC CityGML [11] and OGC IndoorGML [12]. In particular, these standards focus on three-dimensional (3D) geometric information of buildings and defines semantic features such as rooms, walls, roofs, and corridors [4]. However, IFC and OGC CityGML are suitable for general purpose in a wide range of indoor applications, but they lack for the purpose of indoor navigation, especially autonomous navigation of mobile robots, due to the absence of topological constructs. In other words, they need additional cost to compute a path from one position to another and guide a route. OGC IndoorGML defines a framework of indoor spatial information of building components and a network model for their topological relationships, e.g., adjacency and connectivity, for indoor navigation. However, it has a limitation to integrate navigation in indoor and outdoor environments. Still, most navigation systems are focused on only outdoor or indoor environment.

In this paper, we consider a spatial model to integrate the indoor and outdoor navigation spaces for seamless path planning. There are several studies for making unified models for indoor-outdoor seamless navigation [8,13,15,16]. Wang and Niu [16] proposed a data model based on the OSM basic data structure for mapping the connections between inner building parts; e.g., rooms, doors, stairs, elevators, etc. Also, they provide a routing algorithm to calculate the indoor-outdoor route using the proposed data model. However, these unified models have a common disadvantage: the cost to convert the existing models into a unified model. Instead of the unified model, we propose a reference model as an extension of IndoorGML to combine the indoor and outdoor features for seamless navigation in this study. In particular, we take into account the interoperability and compatibility of standardized data models and formats. This paper consists of four parts as follow: In Sect. 2, we describe the existing studies about indoorGML based navigation models and spatial data fusion methods. In the next section, we briefly introduce IndoorGML. In Sect. 4, we present the proposed seamless indoor-outdoor navigation model based on IndoorGML. In the next section, an experiment on the proposed model with CityGML is described, and the final section summarizes of this study and areas of further research.

---

[1] https://www.google.com/maps/about/partners/indoormaps.

[2] https://wiki.openstreetmap.org/wiki/Indoor_Mapping.

[3] https://register.apple.com/resources/imdf.

2   Related Work

Since IndoorGML has been published, various studies have been carried out to support navigation service [1,5,6,14]. Ryu et al. [14] presented Voice-based Indoor Maps (VIM) for visually impaired people. VIM is developed based on IndoorGML with extension model for landmark information of braille block. However, since it is based on IndoorGML, there is a limitation that it can be used only for indoor braille blocks. This is a common limitation of extending models based on IndoorGML.

Also, there are several studies on how to create an integrated model with IndoorGML and other standards [7,10]. Kim et al. [7] presented the methods for integrating IndoorGML and CityGML using the mapping between feature types of those standards. Also, they presented methods and guidelines of the automatic derivation of IndoorGML instance from CityGML LoD dataset. However, this study only considers connections between elements that represent the indoor space of two standard documents.

Lastly, there are a few studies about the connection between indoor and outdoor spaces based on IndoorGML [9,13]. Park et al. [13] presented the Topological Relation-based Data Fusion Model (TRDFM) using topological relations among spatial objects. Similar to our study, TRDFM was developed based on the concept of Anchor node in the IndoorGML document. However, this study extended the representation of outdoor spaces based on elements of IndoorGML. This method can not take advantage of existing standard data for outdoor space, so there is a restriction that TRDFM must rewrite outdoor spatial data even if data is representing the same area. To overcome these issues, we propose an extension model that making the connection between indoor and outdoor spaces, to integrating IndoorGML and other standards.

3   IndoorGML

The OGC IndoorGML was established as the OGC international standard for indoor navigation applications and XML-based formats to represent indoor spatial information. In this section, we briefly introduce IndoorGML and suggest the core concept of the model for indoor-outdoor seamless navigation service.

3.1   Basic Concept

IndoorGML expresses indoor space as two spatial models: Euclidean space represents the shape of cell spaces that is an element of the indoor space; Topology space represents the connectivity and adjacency between cell spaces. An indoor network can be derived from Topology Space using the Poincaré duality. Therefore, IndoorGML utilizes a network model for navigation and expresses the connectivity relationships among cell spaces. The nodes of the indoor network represent rooms, corridors, doors, elevators, and staircases. The edges of the indoor network represent the topological relationships among indoor spatial entities and can indicate the paths of pedestrian movement between nodes within a building. Therefore, one edge should be represented by two nodes. The network model in IndoorGML is represented by nodes (as called **State**) and edges (as called **Transition**) feature, as shown in Fig. 1.

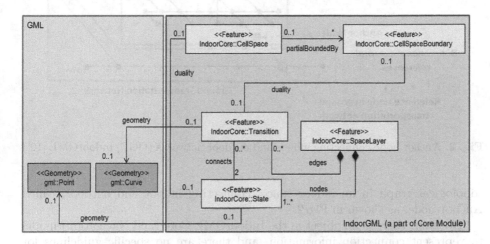

Fig. 1. Part of IndoorGML Core module UML diagram (OGC, IndoorGML [12])

3.2 Thin and Thick Model

IndoorGML is divided into two models depending on the representation of wall (and door) thickness. Thick wall model represents walls (and doors) has a certain thickness. In the case of thick wall model, walls and doors should be represent **CellSpace** (and **State**). On the other hand, thin wall model represents walls (and doors) has no thickness, like paper. In the case of thin wall model, walls and doors should be represent **CellSpaceBoundary** (and **Transition**). We need to consider both models for connection between indoor space and outdoor space. However, when expressing an entrance with Thin wall model, one **State** is required in the outdoor space according to the definition of transition. However, since **State** has a duality relation with **CellSpace** as shown in Fig. 1, it is necessary to express the outdoor space as **CellSpace** in order to create a **State** in outdoor space. However, this is not semantically equivalent to **CellSpace** defined in IndoorGML. In conclusion, the entrance should always be expressed as **State** like the door of the thick wall model.

3.3 Concept of Anchor Node

IndoorGML contains naive ideas for representing indoor and outdoor connections. In IndoorGML, "Entrance" is represented as a special node of the

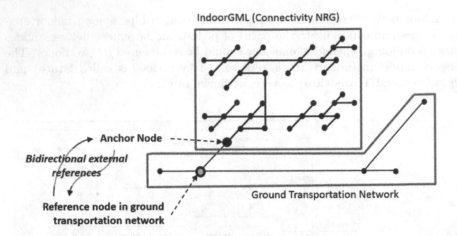

Fig. 2. Anchor node connecting indoor and outdoor networks (OGC, IndoorGML [12])

topological graph in an indoor space, connecting indoor and outdoor, called Anchor node, as shown in Fig. 2.

However, there is no object in the IndoorGML Core (and Navigation) model to represent connection information, and there are no specific guidelines for expressing indoor and outdoor connection information using only existing objects in the IndoorGML schema. To solve this problem, in [13], indoor and outdoor connections are expressed by extending **State** and **Transition** to **SpecialState** (Anchor node) and **SpecialTransition** (Anchor edge) for making an indoor-outdoor unified model. However, this unified model has the problem that all indoor and outdoor data must be described in the document of IndoorGML, even if outdoor data already exists with another format. Also, expressing the outside network using only the elements of IndoorGML do not fit from the characteristic of the elements of IndoorGML. Therefore, we need to extend IndoorGML by expressing entrance elements and making a link for the outside network, for representing indoor-outdoor connection information.

4   Seamless Navigation Model for IndoorGML

This section describes our Seamless Navigation Module (SNM) as an extension of IndoorGML based on the concept of Anchor node, which is the special node that represents the entrance of the building. SNM focuses on reusing other standards that represent outdoor spaces and making connections with them, as shown in Fig. 3. In order to connect two different spaces, we define the following attributes for an instance of Anchor nodes:

- Parameters for CRS conversion.
- External reference to the outdoor transportation network.

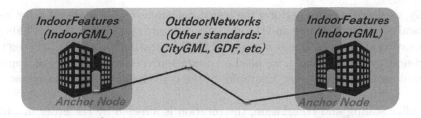

Fig. 3. Concept of Seamless Navigation Model

4.1   Conversion Method of CRS

For seamless navigation, conversion of the Coordinate Reference System (CRS) is necessary to integrate indoor and outdoor coordinate systems. In cases where the global CRS is used for indoor space, the conversion parameters are not necessary. However, in case of using the local CRS, four parameters are required to Cartesian coordinate system conversion:

– origin point of target CRS (or global CRS) $P_o(x_0, y_0, z_0)$,
– re-scaling factor $R(s_x, s_y, s_z)$,
– rotation angles $A(\alpha, \beta, \gamma)$, along x, y, z-axis, and
– translation vector $T(t_x, t_y, t_z)$

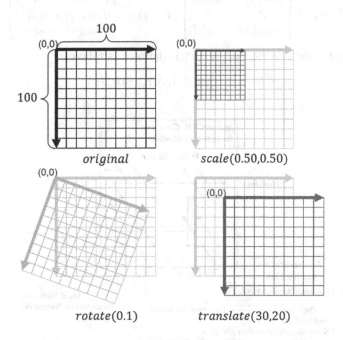

Fig. 4. Example of transform methods (2-dimensional case)

Firstly, we need origin to perform the transformation. Next, a scale value between the local coordinate system and the global coordinate system is required. Thirdly, the rotation angle of each axis is required for the rotation movement between the coordinate systems. Finally, we need a translation vector for parallel movement between coordinate systems. Examples of the conversion methods are shown in Fig. 4.

Unlike scaling and translation, the rotation is affected by the order in which the parameters (rotation angles) are applied. We usually can use Euler angle for 3D rotation, described in [2]. Euler angles are described as a sequence of rotations about three mutually orthogonal coordinate axes fixed in R^3 space. In this paper, we used yaw, pitch, and roll rotation, one of the sequences of Euler angles.

Yaw is a counterclockwise rotation of α around the z-axis. The rotation matrix is given by as follow:

$$R_z(\alpha) = \begin{bmatrix} \cos(\alpha) & -\sin(\alpha) & 0 \\ \sin(\alpha) & \cos(\alpha) & 0 \\ 0 & 0 & 1 \end{bmatrix}$$

Similarly, a pitch is a counterclockwise rotation of β around the y-axis, and a roll is a counterclockwise rotation of γ around the x-axis. The rotation matrix of pitch and roll are given by as follow:

$$R_y(\beta) = \begin{bmatrix} \cos(\beta) & 0 & \sin(\beta) \\ 0 & 1 & 0 \\ -\sin(\beta) & 0 & \cos(\beta) \end{bmatrix}, R_x(\gamma) = \begin{bmatrix} 1 & 0 & 0 \\ 0 & \cos(\gamma) & -\sin(\gamma) \\ 0 & \sin(\gamma) & \cos(\gamma) \end{bmatrix}$$

So, 3D rotation matrix with α, β, γ is defined as follow:

$$R(\alpha, \beta, \gamma) = R_z(\alpha) R_y(\beta) R_x(\gamma)$$

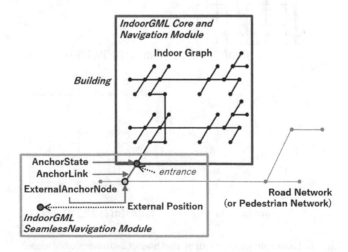

Fig. 5. Concept of IndoorGML Seamless Navigation Model

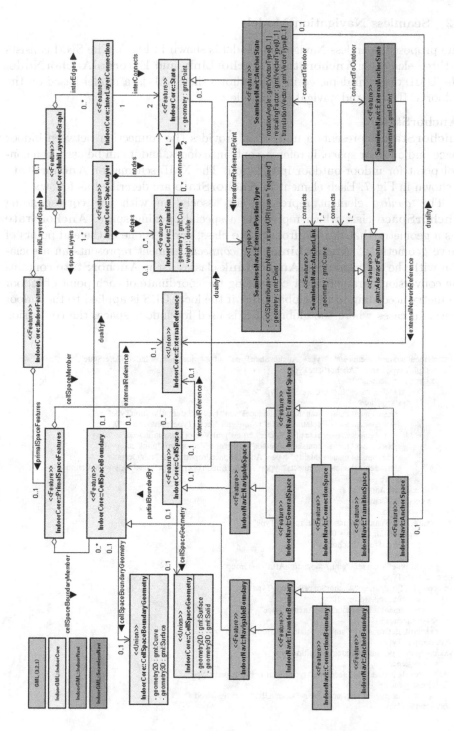

Fig. 6. UML diagram for Seamless Navigation Model

4.2   Seamless Navigation Model

The proposed Seamless Navigation Model is shown in Fig. 5. The SNM consists of three elements: **AnchorState**, **AnchorLink**, and **ExternalAnchorNode**. The UML diagram depicted in Fig. 6 shows the SNM data model based on the IndoorGML core and navigation module.

⟨**AnchorState**⟩
AnchorState represents a node that provides the connection between indoor space and outdoor space. It refers to entrance doors, and it can be used as a control point for indoor-outdoor integration. The XML schema for **AnchorState** is shown in Fig. 7. Each element in **AnchorState** are described as follow:

The *duality* element represents an association with the corresponding **AnchorSpace** class which represents a special opening space. **AnchorState** has a geometry that derived from **State** class, and it is one of the end points of curve geometry of **AnchorLink**. The *connects* element represents an association with the corresponding **AnchorLink** class. Lastly, **AnchorState** contains the conversion parameters for converting the coordinate of each point of indoor geometry according to the global CRS, if the local CRS is applied to the indoor space. In cases where the global CRS is used for indoor space, the conversion

```xml
<xs:element name="AnchorState" type="AnchorStateType" substitutionGroup="IndoorCore:State"/>
<xs:complexType name="AnchorStateType">
    <xs:complexContent>
        <xs:extension base="IndoorCore:StateType">
            <xs:sequence>
                <xs:element name="transformReferencePoint" type="ExternalPositionType"/>
                <xs:element name="rotationAngle" type="gml:VectorType" minOccurs="0"/>
                <xs:element name="rescailingFactor" type="gml:VectorType" minOccurs="0"/>
                <xs:element name="translationVector" type="gml:VectorType" minOccurs="0"/>
                <xs:element name="duality" type="AnchorSpacePropertyType" minOccurs="0"/>
                <xs:element name="connects" type="AnchorLinkPropertyType" maxOccurs="unbounded"/>
            </xs:sequence>
        </xs:extension>
    </xs:complexContent>
</xs:complexType>
<xs:complexType name="AnchorStatePropertyType">
    <xs:sequence minOccurs="0">
        <xs:element ref="AnchorState"/>
    </xs:sequence>
    <xs:attributeGroup ref="gml:AssociationAttributeGroup"/>
</xs:complexType>
<xs:complexType name="AnchorSpacePropertyType">
    <xs:sequence minOccurs="0">
        <xs:element ref="IndoorNavi:AnchorSpace"/>
    </xs:sequence>
    <xs:attributeGroup ref="gml:AssociationAttributeGroup"/>
</xs:complexType>
<xs:complexType name="ExternalPositionType">
    <xs:sequence>
        <xs:element name="geometry" type="gml:PointPropertyType"/>
    </xs:sequence>
    <xs:attribute name="srsName" type="xs:anyURI" use="required"/>
</xs:complexType>
```

Fig. 7. XML schema for **AnchorState**

parameters are not necessary. The *TransformReferencePoint* is **ExternalPosition** type element and a reference point that using for conversion. **ExternalPosition** is a point in the global CRS. Therefore, **ExternalPosition** must have an attribute *crsName* to represent the used outdoor network's CRS.

AnchorState have elements for CRS conversion: *transformReferencePoint* $p_o(x_0, y_0, z_0)$, *rotationAngle* $R(s_x, s_y, s_z)$, *rescailingFactor* $A(\alpha, \beta, \gamma)$, *translationVector* $T(t_x, t_y, t_z)$. Conversions using these parameters will produce results in the order in which they are applied. This paper assumes that the transformation is performed in the order as shown in Fig. 8: *Rotation* → *Scaling* → *Translation*.

In the case of rotation, the rotation is performed after shifting to the origin, based on the **AnchorState** point $p_a(a_x, a_y, a_z)$ for simplification of the problem. Finally, the method to obtain the conversion result,

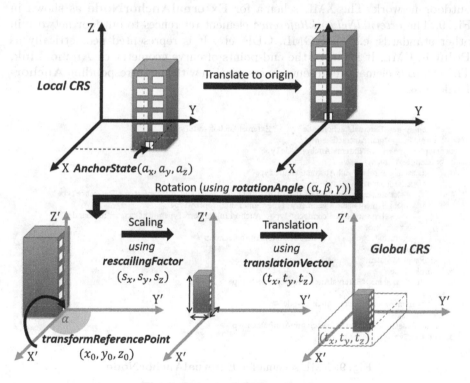

Fig. 8. Process of CRS conversion

$Convert(x, y, z, p_a, p_o, R, S, T)$ using the given parameters is as follows:

$$Convert(x, y, z, p_a, p_o, R, S, T)$$
$$= R_z(\alpha) R_y(\beta) R_x(\gamma) S(x - a_x, y - a_y, z - a_z) + p_o + T$$
$$= \begin{bmatrix} \cos(\alpha)\cos(\beta) & \cos(\alpha)\sin(\beta)\sin(\gamma) - \sin(\alpha)\cos(\gamma) & \cos(\alpha)\sin(\beta)\cos(\gamma) + \sin(\alpha)\sin(\gamma) \\ \sin(\alpha)\cos(\beta) & \sin(\alpha)\sin(\beta)\sin(\gamma) + \cos(\alpha)\cos(\gamma) & \sin(\alpha)\sin(\beta)\cos(\gamma) - \cos(\alpha)\sin(\gamma) \\ -\sin(\beta) & \cos(\beta)\sin(\gamma) & \cos(\beta)\cos(\gamma) \end{bmatrix}$$
$$* \begin{bmatrix} s_x * (x - a_x) \\ s_y * (y - a_y) \\ s_z * (z - a_z) \end{bmatrix} + \begin{bmatrix} x_0 + t_x \\ y_0 + t_y \\ z_0 + t_z \end{bmatrix}$$

⟨**ExternalAnchorNode**⟩

ExternalAnchorNode represents a node that represents the position on the outdoor network. The XML schema for **ExternalAnchorNode** as shown in Fig. 9. The *externalNetworkReference* element references to outdoor network in other standards; e.g., CityGML, GDF, etc. It is represented geometrically as **Point** in GML. It is one of the end points of curve geometry of **AnchorLink**. The *connects* element represents an association with the corresponding **AnchorLink** class.

```
<xs:element name="ExternalAnchorNode" type="ExternalAnchorNodeType"
substitutionGroup="gml:AbstractFeature"/>
<xs:complexType name="ExternalAnchorNodeType">
    <xs:complexContent>
        <xs:extension base="gml:AbstractFeatureType">
            <xs:sequence>
                <xs:element name="externalNetworkReference" type="IndoorCore:ExternalReferenceType"/>
                <xs:element name="geometry" type="gml:PointPropertyType"/>
                <xs:element name="connects" type="AnchorLinkPropertyType" maxOccurs="unbounded"/>
            </xs:sequence>
            <xs:attributeGroup ref="gml:AssociationAttributeGroup"/>
        </xs:extension>
    </xs:complexContent>
</xs:complexType>
<xs:complexType name="ExternalAnchorNodePropertyType">
    <xs:sequence minOccurs="0">
        <xs:element ref="ExternalAnchorNode"/>
    </xs:sequence>
    <xs:attributeGroup ref="gml:AssociationAttributeGroup"/>
</xs:complexType>
```

Fig. 9. XML schema for **ExternalAnchorNode**

⟨**AnchorLink**⟩

AnchorLink represents an edge between the indoor network and outdoor networks. The XML schema for **AnchorLink** as shown in Fig. 10. **AnchorLink** must have connection **AnchorState** and **ExternalAnchorNode**. For the geometrical representation of an **AnchorLink**, a **Curve** geometric primitive object from the GML is used. This geometry is derived from the geometry of **AnchorState** and **ExternalAnchorNode**.

```
<xs:element name="AnchorLink" type="AnchorLinkType" substitutionGroup="gml:AbstractFeature"/>
<xs:complexType name="AnchorLinkType">
    <xs:complexContent>
        <xs:extension base="gml:AbstractFeatureType">
            <xs:sequence>
                <xs:element name="connectToIndoor" type="AnchorStatePropertyType"/>
                <xs:element name="connectToOutdoor" type="ExternalAnchorNodePropertyType"/>
                <xs:element name="geometry" type="gml:CurvePropertyType"/>
            </xs:sequence>
            <xs:attributeGroup ref="gml:AssociationAttributeGroup"/>
        </xs:extension>
    </xs:complexContent>
</xs:complexType>
<xs:complexType name="AnchorLinkPropertyType">
    <xs:sequence minOccurs="0">
        <xs:element ref="AnchorLink"/>
    </xs:sequence>
    <xs:attributeGroup ref="gml:AssociationAttributeGroup"/>
</xs:complexType>
```

Fig. 10. XML schema for **AnchorLink**

5   Experimental Examples

For better understanding, this paper provides a use-case of the SNM for the Transportation model of CityGML 2.0 [11]. This section briefly introduces the CityGML Transportation model and suggests some guidelines for using the sample data to create the SNM data.

5.1   Use-Case with CityGML 2.0

CityGML is an official international standard data model and exchange format to store digital 3D models of cities and landscapes. CityGML support transportation model that focuses on thematic as well as on geometrical/topological aspects, as shown in Fig. 11. The main class is **TransportationComplex**, which is composed of the parts **TrafficArea** and **AuxiliaryTrafficArea**. In the Level of Detail (LOD) 0, the transportation complexes are modeled by line objects establishing a linear network. On this level, pathfinding algorithms or similar analyses can be executed.

Therefore, the geometry of **ExternalAnchorNode** should be created as a point on the *lod0Network* and must have the source data information of the **TransportationComplex** using the GML ID or URL information. Also, the geometry of **AnchorLink** can be derived from two points which are the geometries of **AnchorState** and **ExternalAnchorNode**.

5.2   Case: AIST Tokyo Waterfront Annex

The sample data for the use-case have been conducted at a real site: AIST Tokyo Waterfront Annex in Japan as shown in Fig. 12. This sample data shows the basic structure of the SNM data and how the SNM and CityGML Transportation model datasets are linked via external references. For simplicity, the

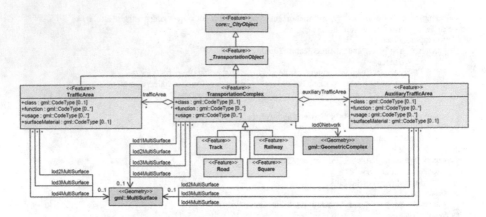

Fig. 11. UML diagram of transportation model in CityGML 2.0 (OGC, CityGML [11])

detailed structure inside the AIST building is not represented, and the data are constructed using only 2D geometry. All geometric data in the sample data are derived from the Open Street Map (OSM) data and have the same CRS; EPSG:4326 (WGS 84). The IndoorGML data consists of two spaces: one **CellSpace** and one **AnchorSpace**. The CityGML data has only one **TransportationComplex** instance.

The detailed contents of sample data for the **AnchorState** class is shown in Fig. 13. This data consists of IndoorGML Core and Navigation modules. **AnchorState** can have a geometry derived from **State** class. This geometry is used to create the geometry of **AnchorLink**. It can have several elements for conversion. However, in this case, all geometries have the same CRS, so these elements are omitted. **AnchorState** must have a *transformReferencePoint* and *connects*. *TransformReferencePoint* must have a CRS information as shown in the yellow part of Fig. 13. *connects* is represented by xlink for GML id of **AnchorLink** class instance, as shown in the green part of Figs. 13 and 16. **AnchorState** can have a *duality* association with the **AnchorSpace** class instance, as shown in the blue part of Fig. 13.

Figure 14 shows **ExternalAnchorNode** sample data. It consists of three properties: *externalNetworkReference*, *geometry*, and *connects*. *ExternalNetworkReference* is a corresponding object in the **TransportationComplex** instance, as shown in the blue part of Figs. 14 and 15. The geometry of **ExternalAnchorNode** is derived from one of the points on a *lod0network*, as shown in the yellow part of Figs. 14 and 15. *connects* is represented by xlink for GML id of **AnchorLink** class instance, as shown in the green part of Figs. 14 and 16.

AnchorLink sample data is shown in Fig. 16. The association elements (*connectToIndoor* and *connectToOutdoor*) are represented by xlinks for GML id of each class instance. The curve geometry is derived from the geometry of *connectToIndoor* and *connectToOutdoor* instance.

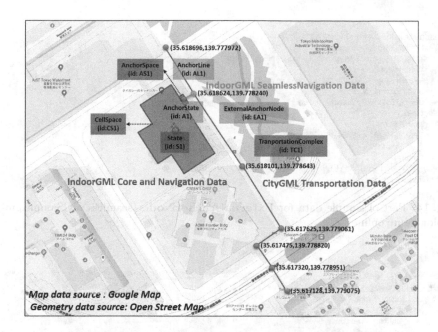

Fig. 12. Use-case site - AIST Tokyo Waterfornt Annex

```
<core:IndoorFeatures gml:id="IFs">
    ...
  <core:cellSpaceMember>
    <navi:AnchorSpace gml:id="AS1">
      <gml:name>Entrance</gml:name>
      ...
    </navi:AnchorSpace>
  </core:cellSpaceMember>
    ...
  <core:stateMember>
    <AnchorState gml:id="A1">
      <core:geometry>
        <gml:Point srsName="EPSG:4326">
          <gml:pos>35.61855174466 139.77813125399</gml:pos>
        </gml:Point>
      </core:geometry>
      <transformReferencePoint srsName="EPSG:4326">
        <geometry>
          <gml:Point srsName="EPSG:4326">
            <gml:pos>35.61855174466 139.77813125399</gml:pos>
          </gml:Point>
        </geometry>
      </transformReferencePoint>
      <duality xlink:href="#AS1"/>
      <connects xlink:href="#AL1"/>
    </AnchorState>
  </core:stateMember>
    ...
</core:IndoorFeatures>
```

Fig. 13. Part of sample data for **AnchorState** (seamlessNaviSample.gml) (Color figure online)

```
<ExternalAnchorState gml:id="EA1">
   <externalNetworkReference>
      <core:informationSystem>cityTransSample.gml</core:informationSystem>
      <core:externalObject>
         <core:name>GMLID_TC1</core:name>
      </core:externalObject>
   </externalNetworkReference>
   <geometry>
      <gml:Point>
         <gml:pos>35.618624 139.778240</gml:pos>
      </gml:Point>
   </geometry>
   <connects xlink:href="#AL1"/>
</ExternalAnchorState>
```

Fig. 14. Part of sample data for **ExternalAnchorNode** (seamlessNaviSample.gml) (Color figure online)

```
<cityObjectMember>
   <tran:TransportationComplex gml:id="TC1">
      <tran:lod0Network>
         <gml:CompositeCurve srsName="EPSG:4326">
            <gml:curveMember>
               <gml:LineString>
                  <gml:pos>35.618696 139.777972</gml:pos>
                  <gml:pos>35.618624 139.778240</gml:pos>
                  <gml:pos>35.618101 139.778643</gml:pos>
                  <gml:pos>35.617625 139.779061</gml:pos>
                  <gml:pos>35.617475 139.778820</gml:pos>
                  <gml:pos>35.617320 139.778951</gml:pos>
               </gml:LineString>
            </gml:curveMember>
         </gml:CompositeCurve>
      </tran:lod0Network>
   </tran:TransportationComplex>
</cityObjectMember>
```

Fig. 15. Part of sample data for **TransportationComplex** (cityTransSample.gml) (Color figure online)

```
<AnchorLink gml:id="AL1">
   <connectToIndoor xlink:href="#A1"/>
   <connectToOutdoor xlink:href="#EA1"/>
   <geometry>
      <gml:LineString srsName="EPSG:4326">
         <gml:posList>
            35.61855174466 139.77813125399
            35.618624 139.778240
         </gml:posList>
      </gml:LineString>
   </geometry>
</AnchorLink>
```

Fig. 16. Part of sample data for **AnchorLink** (seamlessNaviSample.gml) (Color figure online)

6   Conclusion

Indoor and outdoor spatial integration is one of the traditional topics in GIS, particularly location-based services such as navigation for pedestrians (and robots). Several researchers have developed several data fusion methods to integrate the data sets generated based on various data models. This study presented the OGC IndoorGML extension module for seamless navigation between indoor and outdoor space, called Seamless Navigation Model (SNM). The SNM is developed based on the concept of Anchor node. To realize the connectivity of different spatial data based on the topological relations using the SNM, we provided a use-case about connecting with OGC CityGML.

However, further researches are required to implement the purposed extension model to handle the various GIS formats. Further research will include additional use-case with international standards, for example, Geographic Data Files (GDF). And then, finally, we plan to submit the SNM to OGC as a discussion paper with use-cases.

Acknowledgment. This research was partially supported by the New Energy and Industrial Technology Development Organization (NEDO) and a grant (19NSIP-B135746-03) from National Spatial Information Research Program (NSIP) funded by Ministry of Land, Infrastructure and Transport of Korean government.

References

1. Alattas, A., Zlatanova, S., Van Oosterom, P., Chatzinikolaou, E., Lemmen, C., Li, K.J.: Supporting indoor navigation using access rights to spaces based on combined use of IndoorGML and LADM models. ISPRS Int. J. Geo-Inf. **6**(12), 384 (2017)
2. Arfken, G.B., Weber, H.J.: Mathematical methods for physicists. AAPT (1999)
3. BuildingSMART: Industry Foundation Classes (IFC) Version 4.1, IFC4X1 (2017)
4. Chen, J., Clarke, K.C.: Modeling standards and file formats for indoor mapping. In: Procedings of the 3rd International Conference on Geographical Information Systems Theory, Applications and Management, pp. 268–275 (2017)
5. Iida, H., Hiroi, K., Kaji, K., Kawaguchi, N.: A proposal of IndoorGML extended data model for pedestrian-oriented voice navigation system. In: Proceedings of the Seventh ACM SIGSPATIAL International Workshop on Indoor Spatial Awareness, ISA 2015, pp. 2:1–2:6. ACM, New York (2015)
6. Jung, H.J., Lee, J.: Development of an omnidirectional-image-based data model through extending the IndoorGML concept to an indoor patrol service. J. Sens. **2017** (2017). https://www.hindawi.com/journals/js/2017/5379106/cta/
7. Kim, J.-S., Yoo, S.-J., Li, K.-J.: Integrating IndoorGML and CityGML for indoor space. In: Pfoser, D., Li, K.-J. (eds.) W2GIS 2013. LNCS, vol. 8470, pp. 184–196. Springer, Heidelberg (2014). https://doi.org/10.1007/978-3-642-55334-9_12
8. Kim, Y.J., Kang, H.Y., Lee, J.: Development of indoor spatial data model using CityGML ADE. ISPRS-Int. Arch. Photogram. Remote Sens. Spat. Inf. Sci. **1**(2), 41–45 (2013)
9. Lee, J., Kang, H.Y., Kim, Y.J.: Developing data fusion method for indoor space modeling based on IndoorGML core module. J. Korea Spat. Inf. Soc. **22**(2), 31–44 (2014)

10. Liu, L., Zlatanova, S., Zhu, Q., Li, K.: Towards the integration of IndoorGML and IndoorlocationGML for indoor applications. ISPRS Ann. Photogram. Remote Sens. Spat. Inf. Sci. **4**, 343 (2017)
11. Open Geospatial Consortium: OGC City Geography Markup Language (CityGML) Encoding Standard, OGC 12-019 (2012)
12. Open Geospatial Consortium: OGC Indoor Geographic Markup Language (IndoorGML), OGC 14-005r5 (2018)
13. Park, J., Ahn, D., Lee, J.: Development of data fusion method based on topological relationships using IndoorGML core module. J. Sens. **2018** (2018). https://www.hindawi.com/journals/js/2018/4094235/cta/
14. Ryu, H.G., Kim, T., Li, K.J.: Indoor navigation map for visually impaired people. In: Proceedings of the Sixth ACM SIGSPATIAL International Workshop on Indoor Spatial Awareness, ISA 2014, pp. 32–35. ACM, New York (2014)
15. Teo, T.A., Cho, K.H.: BIM-oriented indoor network model for indoor and outdoor combined route planning. Adv. Eng. Inform. **30**(3), 268–282 (2016)
16. Wang, Z., Niu, L.: A data model for using OpenStreetMap to integrate indoor and outdoor route planning. Sensors **18**(7), 1–19 (2018)

A Multi-agent System for the Simulation of Ship Evacuation

Paul Couasnon[1], Quentin de Magnienville[1], Tianzhen Wang[2], and Christophe Claramunt[1,2(✉)]

[1] Naval Academy Research Institute, Lanvéoc-Poulmic, France
christophe.claramunt@gmail.com
[2] Shanghai Maritime University, Shanghai, China

Abstract. Ship evacuation simulations have been used for years to track passengers during an emergency situation. However, current models are still relatively complex to use and are therefore only operated by ship designers at best and rarely by crew members in real situations. The goal of this research is to propose an interactive approach that would be used by crew members on board to organize evacuation drills and be able to interact in real time with the simulation. The approach combines an agent-based model, Python developments and the multi-agent-based Mesa library. Several scenarios of evacuation were tested as well as several pathfinding algorithms. A two-dimensional interface allows different interactive design options such as operating in a given deck, testing different passenger and overload situations. Simulation experiments show promising results as the system developed favors the observation of different evacuation behaviors and outputs as well as highlighting different structural and behavioral patterns in a ship indoor environment.

Keywords: Ship indoor environment · Evacuation simulation · Multi-agent modelling

1 Introduction

In 2006, the International Maritime Organization (IMO) and Maritime Safety Committee (MSC) adopted several amendments to create the concept of Safe Return to Port [1, 2]. These regulations emphasize the preservation of human life on board ships in case of an emergency. They focus on means to evacuate and the time needed for an orderly evacuation. Therefore, ship designers have to run tests to check whether the ship under construction meets the requirements issued by IMO. Two solutions are offered to them. First, they can run an evacuation drill with a crowd on board and count the time taken by every passenger to evacuate. While this is the most direct method to obtain valuable results, the ship must be already constructed. However, this is unpractical and costly to run such a drill and time of evacuation during an exercise can be up to twice as long during a real emergency [3]. The second option is to simulate evacuation, the ship designer only needs the blueprints of various decks of the vessel to use the simulation. This enables to obtain results from a less costly and easier to setup method than the first one. Moreover, such models validated by experimental data have

© Springer Nature Switzerland AG 2019
Y. Kawai et al. (Eds.): W2GIS 2019, LNCS 11474, pp. 63–74, 2019.
https://doi.org/10.1007/978-3-030-17246-6_6

existed for two decades and are now accurate enough to be used by professionals. Nevertheless, evacuation simulations require very precise information about the blueprint of the ship and passengers behaviors. They offer a lot of options that the user can modify but still are very specific and therefore complex to use even for a simple simulation. Importing the geometry of the ship can also be an issue as these programs use specific proprietary formats. The goal of the experimental project presented in this paper is to offer an evacuation simulation software that could be used by crew members on board, with a blueprint importation in a more common file format. Moreover, the options presented to the user should correspond to the ones needed to organize an evacuation drill, such as adding obstacles. This would allow the crew to be able to run simulations handily, and use the software to plan and organize these drills according to situations of interest. To reach that objective, a study has been conducted in three parts. The first one focuses on the theory of evacuation and agent-based model, as well as what already exist in terms of evacuation simulations. Then, the second one describes the specification and the choices made such as the language and libraries used. The third one concerns the simulation design and implementation of the pathfinding algorithms, and finally the visualization interface.

The rest of the paper is organized as follows. Section 2 gives some modelling backgrounds on evacuation models while Sect. 3 introduces the principles of the multi-agent modelling approach. Section 4 develops and reports on the simulations experiments. Finally Sect. 5 concludes the paper and draws some directions for further work.

2   Evacuation Modelling Background

In a landmark behavioral theoretical paper, Kuligowski suggested five behavioral facts to understand how people react to a given emergency event such as a fire in an indoor environment [4]:

- People's first reaction is to feel safe if the fire is not directly close to them. Until they do not understand the situation, they won't rush to exit the place.
- Humans will seek information to understand a given situation. People will ask others if they know what is occurring.
- People act rationally and will help others. They will not panic and will help trapped or injured people.
- Humans will try to take some of their personal belongings. They will try to take clothes and useful items like a coat or a flashlight.
- People will use the paths they are used to. They will move towards familiar exits that they have already identified.

These facts show that passengers will not evacuate efficiently just after the alarm and will delay the beginning of their evacuation. A comprehensive evacuation model should take into account these behaviors. The first goal during an emergency is for the crew to make the passengers aware that there is a real need to evacuate, and that the situation is not a drill. Therefore, evacuation announcements should be made and repeated. Also, crew members with recognizable uniforms should take control of the situation and guide people. Once a few passengers start to evacuate, the snowball effects that will lead the others to evacuate will start [3]. The IMO has implemented several regulations to improve the evacuation of ships during emergencies [2]. The IMO also published a set of guidelines to design ships that are more efficient at mass evacuation [5]. The evacuation time needed depends on the number of passengers and crew members, the geometry of the ship, the type of emergency and the decision process of people on board. An Evacuability Index (Evi) has been proposed to model how fast a ship can be evacuated using the previous inputs. This enables to see on a graphic interface the most used paths as well as the areas of congestion [2]. A study also showed that a 15% to 20% subset of the ship decreases by 33% the walking speed of passengers [6].

Current methods to simulate an evacuation are usually based on either flow-based, cellular automata or agent-based models [7]. Flow-based models use a continuous graph made of nodes and arcs to simulate rooms and the way between them. Node and arc capacities are fixed by the user for the number of passengers allowed. These models are adapted to simulate a crowd flow, but do not take interactions between passengers into account. Cellular automata models use grids made of cells. The discretization of cells allows for each one to have a state that can be modified based on the state of the neighbor cells. These models are computationally straightforward but realism is limited. With agent-based models, each passenger is represented by an agent that has a behavior that allows him/her to move on a grid or on a continuous surface. Agents can have different behaviors and speed, they can interact and communicate with each other. Simulations can be triggered as well as interactive visualizations It is also easy to visualize this model in action, this allowing to supervise the movement of the agents. We decided to retain the capabilities of a multi-agent system and specify them in several complementary dimensions:

(1) The **behavioral model** represent the agents' decisions, that is, actions and movements whose objectives are first to find out exists, avoiding bottlenecks, and also helping other agents whenever possible. Different agent characteristics are taken into account such as being a crew member or an elderly person, this affecting the way they will evacuate and communicate to others.

(2) The **structural model** represents the indoor environment by a regular grid. At a given time, an agent is assigned to one cell and can move step by step, according to its behavior, to other cells step by step. This environment should be partially observable for agents. They should be able to know the position of the exits and see the others passengers and obstacles in the field of view.

(3) The **simulation model**. Different movement behaviors are taken into account, depending on the density of people in the considered space. Every cell of the grid is given a number value corresponding to its distance from the exit. Passengers will move towards cells with lower numbers to find the exit. The speed and flow given to agents are modelled differently at certain key points of the ship. Every agent is given a 360° personal space around them in which no object or other agent can penetrate. The path finding algorithm is based on the implementation of several path finding algorithms, and allows to find the shortest way from one point to another on a grid with obstacles, and allow to give more or less weight to specific areas which means for instance that an area with smoke would be given less weight in the making process for the path than a safe corridor. The simulation

also takes into account the fact that passengers who are evacuating cannot collide. There are various types of collision such as front to front collision, side to side if the passengers go roughly in the same direction and front to back if a passengers needs to overtake another one. To avoid colliding, agents can change their speed, their direction or both. However, they must not get far from the path planned. To decide which passenger will alter his course, some priority rules can be attributed to certain types of agents such as elderly people for instance [8].

3   Multi-agent Modelling Approach

In order to create the multi-agent system, three popular programming languages often used in multi-agent systems have been reviewed, that is, Python, NetLogo and Java/C++. Python has been selected as it offers a good compromise between rapid prototyping and enough modelling capabilities. Not only it is relatively easy to code, it also offers several libraries such as agent-defined, geometrical and temporal primitives, and graphic visualization capabilities at the interface level. Several scenarios of evacuation as well as their combination have been tested (Table 1). The main one is the visualization of the passengers exiting the ship. The other outputs should include the time of evacuation of 90% of the passengers, the time of evacuation of 100% of the passengers and a heat map that shows the areas where passengers passed the most and the eventual bottlenecks. The scenarios have been tested on a real deck boat. MV Glory Sea is a Chinese passenger ship that has a length of 180 m and a capacity of 838 passengers (Fig. 1). Her 5th deck was chosen to test the simulation since it contains both passenger rooms and a restaurant.

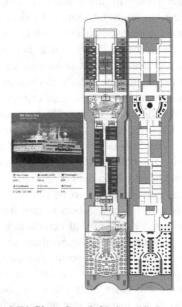

Fig. 1. MV Glory Sea (left), her 5th deck (center)

As illustrated in Table 1 different scenarios have been evaluated.

Table 1. Evacuation scenarios

Scenario	Description	Inputs
Sc.1 Normal	Passengers randomly placed on the deck seeking the nearest exit	Number of passengers
Sc.2 Placed passengers	Passengers placed by the user in a situation of interest seeking the nearest exit	Initial locations of passengers
Sc.3 Obstacles	Obstacles are added to block an exit during the evacuation	Number of passengers Location of the obstacle
Sc.4 Deviant behaviors	Passengers not knowing the exits will try to follow the biggest crowd, and crew members will indicate the right path to lost passengers	Number of passengers Percentage of lost passengers Percentage of crew members

Several types of agents were created to implement the system (Fig. 2). The first class created is *Passenger*, the passengers are the agents that make up the crowd. The initialization defines the attributes of the agent. There are two compulsory attributes that are a unique identity and a multi-agent system related. The others concern the behavior: *Self.behavior* and *Self.goal* adapt the behavior to the type of algorithm used; *Self.alive* indicates whether the passenger is still inside or not. When he is rescued, *self. alive* becomes false. The first method associated to an agent is the movement *Step*. This function is the elementary one as it describes what an agent does during its turn. To simulate the evacuation of a crowd, agents need to be able to get information about other agents. These agents include other passengers, but also walls and exits. For example, a passenger cannot go through a wall while seeking an exit to evacuate. Furthermore, obstacles could appear in the environment to hinder the evacuation of passengers. Eventually, the classes *Exit, Wall* and *Obstacle* inherit from the main *Agent* class. These classes are initialized with a unique as the *Passenger* class, but they indeed do not have the function *Step*. Their only task is to be recognized by passengers. Thus, these agents are only be placed at the beginning of the program and will be staying still through the simulation, except for the obstacles that could be placed while the simulation is running.

The ship indoor environment considered consists in a grid on which the agents are implemented. The Python library Mesa was used to create different types of grids. The most appropriate applied was *the SingleGrid.In*, that is, a special grid where there can only be one agent per cell. Two passengers cannot be in the same place and they have to queue if there are passengers in front of them. Besides, the *SingleGrid.In* has series of basic functions such as placing agents on the map or moving. Specific functions can remove an agent or knowing whether a cell was empty or not. However, a specific class *Map* which inherits of *SingleGridIn* was then created. Five functions were added to make this Map class adapted to the study. These functions concern the movement behavior of passengers on the grid, and what cell to move to. Once the grid was ready to receive the agents, it had to be modeled to have the shape of a ship's deck.

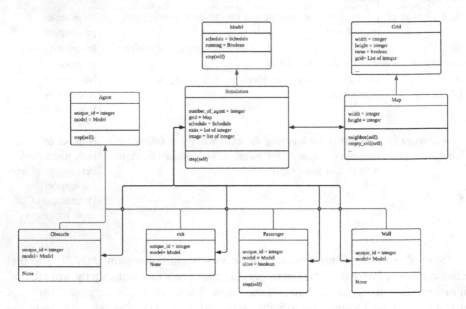

Fig. 2. Agent UML diagram

Different parameters had to be implemented. First, the necessary attributes were *Self.schedule and self.running*. The first one creates a list of agents and activates them in a specific order - here randomly - for them to carry out the action they have to do during their turn. This means that for each step of the program, the order of activation changes. The second one, *Self.running*, is a Boolean acting as a security. If it is on *False*, the steps cannot execute themselves automatically, each one has to be run manually. This prevents the program to run in an infinite loop. It can be set to *True* when program for sure expected to terminate. Then, other attributes needed to be added to meet the requirements of an evacuation simulation. *Self.num_agents* represents the number of passengers on board at the beginning of the simulation. It can easily be changed by the user to the number of passengers required for a given test, this ensuring a sort of scalability under some realistic conditions. *Self.exit* is a list with the location of all exits existing on the grid. Thanks to this list, passengers are able to know the location of exits, like they would in a real situation. The next step was implementing the environment into the program and placing the static agents - walls, exits and forbidden positions - in it. Using the image importing function, an array is created with the same dimensions and with the same number of cells than pixels in the picture designed. After that, these static agents were implemented in the grid with the help of the array.

Eventually, passengers needed to be initialized. First, they have to be placed in the schedule that will activate them when it is their turn to do a step. Then, they have to be placed in the environment with a random position. Thus, the *Random* library was used to give each passenger an unoccupied position. Once the initialization done, the simulation is ready to be run. To do so, the simulation needs a method named *Step* that will apply the *Agent* class step method for every agent in the schedule list.

4   Simulation Experiments

4.1   Graphic Interface Principles

Mesa provides a visualization library to display a background grid and some important values or graphs. It uses a Web server and the programming language Javascript to create and display additional data. The necessary files were already embedded in the visualization module (Fig. 3). A new file Server had to be created to set up the visualization and to contain the portrayal of every agent. A slider on the interface increases or decreases the number of passengers in the simulation interface. The function *Agent_portrayal* depicts the agents during the simulation. It gives agents a different shape or color depending on their type. The function *Visu_agent* calculates the number of agents which are still not rescued. It returns a percentage of this value using the initial number of passengers. Besides, in the *Simulation* class, the attribute *Self.data-collector* was added to collect the result of the function every step. Finally, in its step the model must collect the data by calling this attribute. Eventually, agents had to be implemented on the grid. The method *Create_grid* runs through every cell to analyze what agent needs to be set and what color to fill the tile.

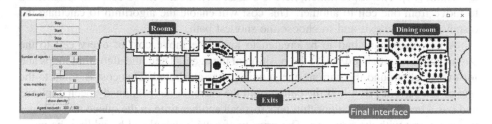

Fig. 3. Simulation interface

4.2   Path Finding Algorithms

Different behaviors were tested to observe how passengers act. Starting from a basic behavior consisting in random walking, several path finding algorithms were implemented and applied. Before creating the behaviors, a method had to be developed to know when an agent is safe. Indeed, the goal of the simulation is to evacuate all passengers and for the simulation to know when an agent has to be removed because it is out of danger. The *Safe* method tests if an agent is rescued, that is to say, being one cell away from an exit cell. First, the method checks if the agent is on board or not, then it analyzes the empty neighboring cells around it to find those that are exit cells. If an exit is found, the *Self.alive* attribute of the passenger class becomes *False* and the agent is removed from the simulation. Therefore, the multi-agent system was able to detect rescued agents and to stop the simulation when all agents were rescued.

The main problem of the Breadth First Search algorithm is that it explores all directions around an agent, whatever the position of the exit [9]. However, in real life a real situation, a passenger knows what direction to go toward the exit since he has at least basic knowledge of the ship. Thus, passengers should focus their search in the direction of the nearest exit. First, a heuristic function that calculates the range from one position to another ignoring walls and obstacles is needed. Then, passengers have access to a list of the exits and the method *Range_exit* creates a realistic solution which represents the shortest range from a passenger to an exit. Then, it loops to calculate each range from the agent's position to the exits with the heuristic function and keeps the best range in the variable. Eventually, the method returns the position of the closest exit. The algorithm Greedy Best First Search uses a similar algorithm to Breadth First Search but a different type of queue is required. In fact, here the top of the queue is the not the oldest element but the nearest from the exit. The *PriorityQueue* class is a list where every cell has a weight, and the method get returns the element with the strongest weight. The closer from the exit a cell is, the heavier its weight is. Instead of returning the first element of the queue, this algorithm will only use the prioritized elements. Therefore, only one direction is explored by the algorithm and not every cell around the passenger, which results in a much faster search.

However, the path found by the Greedy Best First Search algorithm is not the always the shortest. In fact, the algorithm only uses the nearest cell from the exit but the heuristic method used does not take into account walls that could make the way longer. Therefore, the algorithm used had to evolve into a more complex one which is the A* algorithm [10]. The improvement here is that an incremental move cost is implemented to move from one cell to another. This cost will enable the algorithm to calculate the distance from the exit and to choose the shortest path.

A new dictionary had to be created to know the move cost of every cell to the passenger. It is initialized at 0 at the position of the agent. Then, the algorithm loops and for every cell around the current one, a new cost is attributed. This cost is equal to the one of the current cell +1. If the cell around the current one has already been explored and has a lower cost, the algorithm puts this cell another time in the queue and changes its actual cost with the new one. The priority of cells is calculated by adding the move cost from the agent to the cell and the heuristic from the cell to the exit. The dictionary returned at the end delivers the shortest path to the exit. Thus, by improving the Breadth First Search algorithm into the A* algorithm, it was possible to find the shortest path to the exit for every passenger by knowing in which direction to look for at the beginning.

Pathfinding with A* was faster than with Breadth First Search, however a faster method was still sought. Therefore, a potential-based algorithm was tested. With this method, each cell has a number which indicates how many cells are left to reach the exit. Thus, agents do not have to calculate each time the shortest path but only follow the cells in descending order. To create the potential-based algorithm, the Breadth First Search algorithm was used again, not by agents but by the simulation itself. In fact, after creating the grid, the simulation uses the potential method to attribute a number to each cell.

The algorithm works with the same principle as *Breadth_first_search*, except that it starts from the exits and it does not search for a special cell but explore the grid as a whole. A new dictionary labelled *Cost_so_far* pairs each cell with a number. The number 0 is paired with exits, 1 is paired with cells next to exit and so on. On the final grid, each cell has a potential and agents have access to these values. Therefore, each turn they move to a cell with a lower potential and it will eventually lead them to the nearest exit. This method is the fastest since Breadth First Search algorithm is used only once, when the simulation creates the grid. Each turn, agents only have to look for the neighbor cell with the lowest potential which is faster than running the A* algorithm.

4.3 Simulation Experiments

4.3.1 Standard Scenarios

The passengers are not always randomly located on the deck. For example, at dinner time, most passengers are in the dining room and it could be interesting to perform an evacuation from that situation. That is why the function to place agents by the user was created. On the deck at the top, most of the passengers are placed in the dining room. Passengers are randomly placed on the deck and they have to seek the nearest exit. They use the potential method to move and avoid other passengers.

Then, the evacuation starts and agents move to the nearest exit. The most taken paths are showed with the heat map. In most simulations it appears that most of the agents reach the South East exit while the North East exit is unused. Such simulation allows users to observe the global behavior of a crowd on the deck. The heat map shows the most taken paths as well as the bottlenecks that can form represented. One improvement could be to indicate passengers from the northern part of the dining room that there is another exit further than the south east exit but which is not crowded. Thus, improvements could be done to the deck or to the signalization by using these specific simulations.

4.3.2 Scenario with Obstacles

It appears that bottlenecks often form when an exit is blocked by an obstacle. To simulate a real evacuation, some passageways or exits could be blocked with obstacles. In the example presented in Fig. 3, the North West exit is blocked (Fig. 4). Passengers go to the south west exit and a bottle neck is created. Thus, this scenario allows users to predict the situation in the case where paths are blocked by some obstacles.

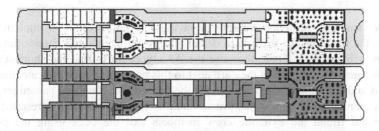

Fig. 4. Evacuation and heat map with the obstacle scenario

4.3.3 Scenario with Deviant Behaviors

Among people on board, several must have a different behavior to have a more realistic simulation. Some passengers do not follow the path to the exit because they do not know where it is located, therefore they will follow the crowd. On order to implement that, deviant agents move to try to reach the center of mass of the agents around them. A range is defined to know which passengers have to be taken into account by the lost one. Then, this agent calculates the center of mass of the passengers in the area defined, and moves in its direction. Thus, the behavior of those agents is unpredictable and they can be blocked inside. To rescue those passengers, crew members need to help them. They are represented in blue in the simulation. To rescue these agents, crew members must seek them to reroute them to the path to the nearest exit. Crew members are represented in green. On the right of the deck, two crew members are reaching two deviant passengers to rescue them (Fig. 5). This simulation allows users to predict how passengers can react and therefore allows us to determine the number of crew members a deck needs to conduct a proper evacuation in function of the percentage of deviant passengers they estimate. It can depend on their age, their knowledge of the ship or their ability to react to an emergency.

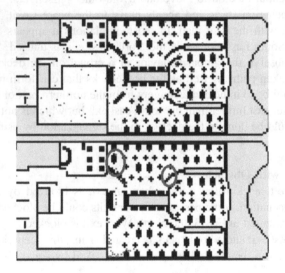

Fig. 5. Evacuation and heat map with the deviant behavior scenario (Color figure online)

Crew members' behavior consists in saving lost passengers. To implement this behavior, the first step for them is to detect all deviant agents and choose the nearest one from their position. Then, they use the A* algorithm to find the shortest path to reach this agent. Finally, when they are next to this passenger, they indicate him where the exit is and the deviant agent will reach the nearest like a normal passenger would. The crew members will repeat this process until every deviant agent is rescued. When everyone has found the exit, the crew members exit the deck using the potential method.

5   Conclusion

The research presented in this paper introduces an experimental ship evacuation modelling approach whose objective is to provide a flexible and experimental simulation environment that can be used by crowd members or even at the design stage. While most current ship evacuation systems are used at the design stage, the peculiarity of our system is that it can be used on the fly by crowd members in a ship. Overall a working simulation environment has been eventually implemented and evaluated with both common and very specific situations. An agent-based model was created with Mesa and enabled agents to have their own behavior. Three pathfinding methods were implemented. Each one was used in a different case in order to be the most efficiently used. Several behaviors such as crew members, disoriented passengers and "normal" passengers were coded, interacting with each other in order to make a realistic crowd model. The main output of the simulation is a visualization interface that provides the user a step by step view of the passengers evacuating the deck. Through a combination of several scenarios, several situations and scenarios can be simulated.

However, features could be added to improve the simulations. For instance, a timer would have allowed to measure the time that the passengers take to evacuate in order to compare it with a real evacuation of the deck. It was not added so far because each step does not take the same time to be executed. The first ones are much longer to execute than the last ones resulting in the time not being proportional to the number of steps. Therefore, a method is yet to be found to resolve this issue since at the moment only the number of steps needed for the evacuation is displayed. Another improvement could be passing from a discrete grid to a continuous environment. That would enable more precise movements and avoidance between passengers.

This prototype can be used by crew members, in particular the person in charge of safety exercises. He would be able to create a scenario and simulate it before having to organize it on board. Moreover, it would allow him to know where to ideally place the crew members for specific situations if the simulation reveals a flaw in the design of the ship or in the way the evacuation is organized. To spot such flaws, the heat map provided at the end of the simulation shows the areas where passengers passed the most, where the bottlenecks were created and the exits that where too much or not enough used. Red spots on the map indicate that the evacuation could be improved at these specific areas.

Modeling several decks of the ship would allow a more global vision of the evacuation. For instance, in some cases it would be shorter for the passenger to use the stairs to go to another deck to find an exit. Moreover, simulating new and more complex behaviors would enable to simulate more complex situations. For instance, while simulating a fire as an obstacle, firefighting crew members could get dressed as such in the right place and then fight the fire to extinguish or isolate it. On a larger scale, this software could enable to understand and apprehend crowd movement and evacuation phenomena in situations other than the emergency evacuation of a ship's deck.

References

1. Maritime Safety Committee: Adoption of amendments to the international convention for the safety of life at sea, 1974, as amended. Resolution MSC, 216:82 (2006)
2. Vassalos, D., Azzi, C., Pennycott, A.: Crisis management onboard passenger ships. In: Human Performance at Sea, Glasgow, pp. 1549–1556 (2010)
3. Boer, L.C., Skjong, R.: Emergency evacuation: how better interior design can improve passenger flow. In: Cruise and Ferry, London (2001)
4. Kuligowski, E.D., Gwynne, S.M.V.: The need for behavioral theory in evacuation modeling. In: Klingsch, W., Rogsch, C., Schadschneider, A., Schreckenberg, M. (eds.) Pedestrian and Evacuation Dynamics, pp. 721–732. Springer, Heidelberg (2008). https://doi.org/10.1007/978-3-642-04504-2_70
5. London IMO: Interim guidelines for evacuation analyses for new and existing passenger ships. MSC/Circ. 1033 (2002)
6. Bles, W., Nooy, S.A.E., Boer, L.C.: Influence of ship listing and ship motion on walking speed. In: Schreckenberg, M., Sharma, S.D. (eds.) Pedestrian and Evacuation Dynamics, pp. 437–452. Springer, Heidelberg (2001)
7. Santos, G., Aguirre, B.: A critical review of emergency evacuation simulation models. Crit. Rev. **1032**, 25–50 (2004)
8. Foudil, C., Noureddine, D., Sanza, C., Duthen, Y.: Path finding and collision avoidance in crowd simulation. J. Comput. Inf. Technol. **17**(3), 217–228 (2009)
9. Moore, E.F.: The shortest path through a maze. In: Proceedings of the International Symposium on the Theory of Switching, pp. 285–292. Harvard University Press (1959)
10. Hart, P.E., Nilsson, N.J., Raphael, B.: A formal basis for the heuristic determination of minimum cost paths. IEEE Trans. Syst. Sci. Cybern. **4**(2), 100–107 (1968)

What Is the Impact of On-street Parking Information for Drivers?

Fabian Bock[1], Sergio Di Martino[2(✉)], and Monika Sester[1]

[1] Institute of Cartography and Geoinformatics, Leibniz University,
Hannover, Germany
{fabian.bock,monika.sester}@ikg.uni-hannover.de
[2] Department of Electrical and Telecommunications Engineering,
University of Naples Federico II, Naples, Italy
sergio.dimartino@unina.it

Abstract. Parking Guidance and Information (PGI) solutions are a well-known class of Intelligent Transportation Systems meant to support drivers by recommending locations and routes with higher chance to find parking. However, the relevance of such systems for on-street parking spaces is barely studied. In this paper, we investigate the consequences of providing the drivers with different parking information to the search. Based on real-world parking data from San Francisco, we investigated the scenario in which a driver does not find a parking space at the destination and has to decide on the next road to go. We consider three different scenarios: (I) No parking availability information; (II) static information about the capacity of a road segment and temporary parking limitations; (III) real-time information collected from stationary sensors. Clearly the latter has strong implications in terms of deployment and operational costs. These scenarios lead to three different guidance strategies for a PGI system. The empirical experiments we conducted on real on-street parking data from San Francisco show that there is a significant reduction of parking search with more informed strategies, and that the use of real-time information offers only a limited improvement over static one.

1 Introduction

Finding an available parking space in crowded urban areas is a pain point of mobility. Even if in the literature there is no clear agreement on the actual extent of this, all the studies point out that parking search is a relevant source of traffic and a relief is overdue [1,2].

A root cause of this problem is that drivers have no knowledge about where there could be a free parking space matching their expectations, and so they have to roam. A special class of Intelligent Transportation Systems, namely the *Parking Guidance and Information* (PGI) systems, might significantly reduce this problem by guiding drivers directly to an area (or a parking facility) with a high likelihood of free spaces, given up-to-date information about the state of parking infrastructure [3].

© Springer Nature Switzerland AG 2019
Y. Kawai et al. (Eds.): W2GIS 2019, LNCS 11474, pp. 75–84, 2019.
https://doi.org/10.1007/978-3-030-17246-6_7

Many studies claim that, thanks to PGIs, the probability of finding available parking spaces is significantly increased [4], while the search time is reduced [5]. Nevertheless, the actual wide-scale adoption of PGIs is limited by difficulties in collecting detailed real-time parking availability information, especially for on-street parking [3,6,7]. Broadly speaking, there are three main typologies of on-street parking information which could be useful to be fed to a PGI system:

1. *Static Parking Information*, containing details about the location of on-street parking lanes, together with their vehicle capacity and optional temporal restrictions. These information are usually provided by the city authorities. However, as maps of parking lanes are often not available, manual recording or crowd-sensing solutions are sometimes needed to map parking lanes [8,9]. Parking capacity can change over time of day both for regular reasons, like peak-hour driving lanes and street cleaning, and on a irregular basis for e.g. construction sites. If each parking space is equipped with a parking meter, the capacity of the lane is fixed. Else, the capacity depends not only on the length of the lane but also on the length of the cars and their distance to each other.
2. *Dynamic Parking Availability Information*, i.e. dynamic information about the most recent measurements of the parking availability in every road segment. An approach to obtain this information is to deploy a multitude of static sensors in the parking infrastructure, like magnetometers in the asphalt [10] or cameras on poles overseeing parking lanes [2]. This solution usually leads to high quality parking information, collected in almost real-time, but it is very expensive to install and maintain on a city-wide scale [6,11]. Another strategy exploits participatory or opportunistic crowd-sensing [12,13] via mobile apps [3] or probe vehicles [14]. Opportunistic mobile apps are cheap to deploy, but they need very high penetration rates to obtain an adequate amount of parking availability information [15]. Probe vehicles can represent an advantageous trade-off between deployment costs and coverage of on-street parking monitoring, but their deployment requires a clear commitment of many mobility stakeholders, and to date is not widely employed [7]. Moreover, with crowd-sensing strategies, there could be a consistent time gap since the last availability observation on a given road segment. Indeed, in [7], authors reported an average interpolling time per segment, in San Francisco, ranging from less than 4 min up to hours, when using about 500 taxis as probes.
3. *Parking Availability Predictions*, i.e. highly dynamic information intended as predictions of the parking availability at arrival time, based on historical parking observations [16]. This scenario is not considered in the rest of the paper.

While in general these solutions to feed PGIs are complex and/or expensive to deploy, the actual benefits for the driver of real-time parking availability information in PGIs have been barely studied, and to the best of our knowledge, there are only investigations based on simulated data [17,18]. For the search routing without additional information, simulator studies investigated the drivers' strategies to find parking [19]. To estimate the amount of parking search based on real-world sensor data, Millard-Ball et al. [20] simulated a random drive through the road network and measured the search length until the first free spot was found. In [18], authors simulated the potential savings of PGIs, but limiting to off-street parking. The benefit of parking availability information exchanged among vehicles and with the infrastructure is investigated in [17], where authors found that PGIs fed by static sensors can reduce parking search time by up to around 50% in highly occupied scenarios, while no relevant reductions can be found in case intelligent vehicles exchange parking and leaving notifications without the support of additional sensors.

Thus, to the best of our knowledge, there exists a lack in the literature, since no studies analysed in an empirical way the actual benefits of deploying such a costly parking sensing infrastructures to feed on-street parking availability information to PGIs.

To address this open issue, in this paper we present the preliminary results of a study based on real on-street parking data, investigating the potential benefits of different kinds of information that can be provided to PGIs, in terms of inducted reduction of on-street parking search. In particular, by exploiting 2 weeks of real parking availability data from the SFPark project in San Francisco (USA), we compared the effects of the following scenarios for the drivers: (I) no parking information at all; (II) only static information about the parking capacity of the nearby road segments, plus temporary parking limitations; (III) real-time parking availability information collected from stationary sensors.

The main contributions of this paper to the body of knowledge are the definition of a representative scenario suitable for evaluating the benefits of different parking information to be used in PGIs, and the analytical comparison of four parking guidance strategies, based on real-world data, in the Municipality of San Francisco.

2   Experimental Setup

The focus of the present research is to investigate the influence on parking search of different guidance strategies for a PGI, that can be exploited on top of different kinds of on-street parking information. In this section we describe the experimental set-up of our empirical study. In particular we start by describing the real-world dataset we used, and then on top of this, we present the investigated research question and used data sources.

2.1   The On-street Parking Availability Dataset

For the experimental evaluation, we used public data sources related to mobility in San Francisco (USA). Information about the topology and characteristics of the road network have been extracted from OpenStreetMap. For the considered area of investigation, we have more than 500 road segments, of which about 40%

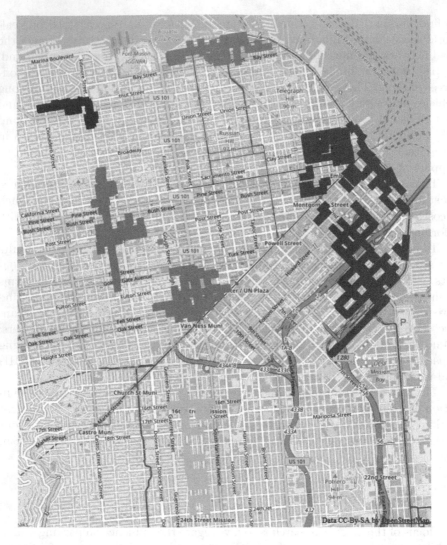

Fig. 1. Map of the sensor-equipped road segments in San Francisco. Background map is taken from *OpenStreetMap*.

are one-way roads. On average, for each road segment there are 2.7 manoeuvre options at the subsequent intersection.

The parking data stems from one of the largest on-street parking projects worldwide, called *SFpark* [10]. In this project, a total budget of $50 million dollars was spent to investigate demand-responsive pricing of on-street parking. A main part of the project was the equipment of about 8,000 parking spaces with static sensors in six districts (see Fig. 1). Even though 8,000 equipped stalls is a remarkable number, this is less than 3% of the total number of on-street legal parking spaces in San Francisco [10]. These numbers make clear the problems and the costs to scale the instrumentation of on-street parking stalls to a city-wide dimension.

These sensors continuously monitored the parking availability and transmitted the status to a central instance. The availability data per block face was provided to the public via a smartphone application and a data API. In this study, we used a record of this API data in a 5-min interval from June 2013. Every record contained the number of available parking spaces and the total parking capacity per block face.

Since SFpark reported several sensor problems [10], we applied some cleansing to avoid implausible data. We omitted data for all segments where there was no data or a constant availability for more than three days. Also, segments were filtered out where parking occupancy rate was always lower than 85% during the two-week period. In total, 420 block faces remained after filtering.

In nearly 10% of time steps and road segments (also called *cases* in the following), parking is not allowed. When parking is allowed, 8 parking spaces is the most frequent capacity. Also, the capacity is very heterogeneous with values between 1 and 36. In the entire dataset, only in about 8% of all cases when parking is allowed, all spaces were occupied and thus an hypothetical driver had to roam and take a decision at the next intersection. This is a quite surprising number, far from common experience in many other dense urban scenario around the world. From these numbers, it seems that the areas covered by the SFpark project are not so problematic, in terms of lack of free parking spaces, as also found by Millard-Ball et al. [20].

2.2 Research Question

The availability of parking information becomes crucial when the driver does not find a parking space in the road segment[1] of his/her destination, as illustrated in Fig. 2, and starts to roam. In this situation he/she needs to be supported by a Parking Guidance and Information system, to decide which road segment to go next. Such a PGI can implement different guidance strategies, depending on the kind of available information. In particular, we considered the following strategies:

1. *No Information:* No PGI at all. The drivers do not have access to any parking information and just randomly choose the subsequent road segment. This represents for us the **baseline**.
2. *Static Parking Legal Information:* the PGI excludes from the guidance a road segment with parking prohibitions, randomly choosing among the others at the intersection.
3. *Static Parking Legal and Capacity Information:* the PGI suggests to the driver to turn on the next road segment with highest legal parking capacity.
4. *Actual Parking Availability:* the PGI suggests to the driver to turn on the road segment where the sensed parking availability is highest.

---

[1] In the following we will identify with the term *road segment* the part of a road between two subsequent intersections.

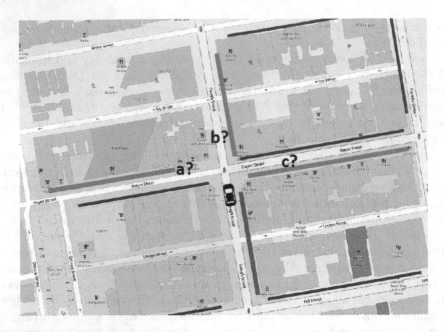

Fig. 2. The investigated scenario, where a driver does not find a parking space at the destination (parking lane in red) and has to choose a road segment at the subsequent intersection. (Color figure online)

In our empirical evaluation, using real data of on-street parking availability, we simulated the scenario where a driver does not find a parking space on the road segment of his/her destination (among one of those covered by the Sfpark sensors), and thus requires recommendations by a PGI that is based on one of the four above described strategies. Then, we measured the number of **unsuccessful decisions**, i.e. not finding a parking in the next road segment (labelled with a, b and c in Fig. 2) according to real SFpark data, for the three different guidance strategies plus the baseline.

This means that we investigated the parking search to the next road segment, since, as done in similar works based on simulated data, this search can be seen as a series of en-route decisions at the intersections [21], where, if there is also no available space at the next road segment, the driver has to decide for a following segment again[2].

There are some assumptions to state, to measure their impact of the different guidance strategies. Following the parking search simulation in [20], we assume that a driver could take any road segment connected to the current one, while respecting regulations of one-way roads (in contrast to [20]), following the PGI driving direction.

Given all the time frames of 5 min over the 2 weeks of investigation in the dataset, we analysed only segments without available parking spaces at a given

---

[2] Please note that we neglect walking distances in this consideration.

time, to simulate the scenario of Fig. 2, where a driver arriving at this road segment needs to decide for a next segment at the subsequent intersection. We count the ratio of unsuccessful decisions, meaning that there is no parking also in the subsequently chosen road segment. Moreover, parking guidance is only able to help at intersections where there is at least one subsequent road segment with a free parking space. Also, if there is parking in all subsequent road segments there is no need for a parking guidance system. Finally, we assume that drivers do not take U-turns. As discussed in [20], this maneuver is illegal in some cases but still many drivers do it. Thus, the resulting numbers describing the parking search phenomenon represent a rather lower bound of the problem.

Given these assumptions, we performed an evaluation of unsuccessful decisions for all the road segments without available parking spaces, regardless of the availability in the next segments.

Thus, our research question was: *What is the impact of different Parking Guidance strategies for on-street parking search?*

3 Results and Discussion

In this section we report the results of the empirical study, evaluated by considering all time steps and road segments of the dataset described in Sect. 2.1, where no parking space is available.

In the evaluation period of two weeks, over than about 1.5 million availability observations, there were 95,403 cases in which all parking spaces of a road segment were occupied.

Table 1. Rate of unsuccessful choices for complete dataset

No information	16.0%
Static information (only parking prohibition)	11.8%
Static information (prohibition + lane capacity)	5.6%
Availability from sensors	3.2%

The results for the different guidance strategies are reported in Table 1. More in details, if the subsequent road segment is chosen randomly, the driver does not find parking also in that road segment in about 16% of the cases. If information on parking prohibitions are considered for the guidance, the ratio of unsuccessful cases is reduced to about 12%. With a guidance strategy exploiting the parking capacity information of road segments, the ratio of unsuccessful choices drops to only 5.56%. By introducing real-time data, the ratio becomes 3.17%.

If we consider relative reductions, failures in finding parking in the subsequent road segment are reduced by up to 79% using a parking guidance strategy based on real-time data, compared to no information at all.

3.1 Discussion

The achieved reduction of unsuccessful turn choices by up to 79% is a considerably high percentage. It also means that parking search traffic could be reduced by up to 79% if we assume that drivers always make a new decision at every intersection. This assumption is a typical search behavior according to [21] and is also used in parking search simulations like [19]. It only neglects that drivers might also consider the walking distance to the destination. Drivers might not take any possible free parking space they find, e.g. due to the walking distance or an upcoming tow-away periods. Also, some drivers start to search for parking before reaching the destination, which is not considered in our settings.

Already the guidance based on the capacity information strongly reduces the errors by more than 60% for the full SFpark investigation area. This strong result implies that in such scenarios with high capacity heterogeneity there might be no need for systems that obtain parking availability information. Nevertheless, for routing decision with less heterogeneous capacity, the capacity layer becomes irrelevant but all cases for parking availability information still provided very reliable route suggestions close to the optimum. Thus, the relevance of parking availability information, either from static sensors or probe vehicles, appears to be strongly depending on the heterogeneity of the parking capacity in the city district.

The potential reductions in parking search are higher in our results compared to the simulation study in [17] where reductions were less than 50% even in the most occupied settings. A main difference of their model to our real-world data is the heterogeneity in parking capacity and demand. Their model assumed a homogeneous parking demand except for one city block and a homogeneous capacity of the parking lanes throughout the network. It is also discussed in their study that a higher heterogeneity might further reduce the parking search time using a parking guidance system. However, we only considered one vehicle on the search while their simulation contained several vehicles.

The number of informed vehicles searching at the same time presumably also affects the benefits of a parking guidance system. Especially, if only the capacity layer is used all drivers would go to the same road segments regardless of the actual availability. This could lead to the inverse situation that the road segments with the highest parking capacity would have the least availability in some cases.

4 Conclusion

With the advent of on-line vehicles, Parking Guidance and Information systems are supposed to become more and more diffused, being supposed to be a great help for drivers searching a parking space. Nevertheless, to the best of our knowledge, no studies have been conducted to assess their effectiveness on the search for an on-street parking space, using real data.

In this paper, we empirically investigated the potential benefit of different layers of a dynamic parking map that can be used in a parking guidance system.

Based on real-world data from San Francisco, we evaluated these information in a scenario where drivers did not find parking at the destination and start to search. We found that the chance of ending up in a fully parking road segment on the next block can be reduced by up to 79%. In particular, the capacity of parking lanes is already a very important hint for the search if the capacity is heterogeneous in the area. For the case that the parking capacity is more homogeneous, parking availability information strongly reduces the cases of unsuccessful search.

There are many future evolutions of this preliminary research. It would be interesting to consider also crowd-sensed parking availability information, whose quality is lower than the one coming from static sensors, with important implications on PGIs. We would also extend the scenario from one following road segment to a full search route computed based on the information in the dynamic parking map. Finally, it would also be interesting to investigate the benefits of parking guidance systems based on real-world data from other cities with different layout and more significant parking problems.

Acknowledgment. This research has been supported by the German Research Foundation (DFG) through the Research Training Group SocialCars (GRK 1931). The focus of the SocialCars Research Training Group is on significantly improving the city's future road traffic, through cooperative approaches. This support is gratefully acknowledged.

References

1. Shoup, D.C.: Cruising for parking. Transp. Policy **13**(6), 479–486 (2006)
2. Lin, T., Rivano, H., Mouël, F.L.: A survey of smart parking solutions. IEEE Trans. Intell. Transp. Syst. **PP**(99), 1–25 (2017)
3. Ma, S., Wolfson, O., Xu, B.: UPDetector: sensing parking/unparking activities using smartphones. In: Proceedings of the 7th ACM SIGSPATIAL International Workshop on Computational Transportation Science, pp. 76–85. ACM (2014)
4. Teodorović, D., Lučić, P.: Intelligent parking systems. Eur. J. Oper. Res. **175**(3), 1666–1681 (2006)
5. Axhausen, K., Polak, J., Boltze, M., Puzicha, J.: Effectiveness of the parking guidance information system in Frankfurt am Main. Traffic Eng.+ Control **35**(5), 304–309 (1994)
6. Xu, B., Wolfson, O., Yang, J., Stenneth, L., Yu, P.S., Nelson, P.C.: Real-time street parking availability estimation. In: 2013 IEEE 14th International Conference on Mobile Data Management (MDM), vol. 1. IEEE, pp. 16–25 (2013)
7. Bock, F., Di Martino, S.: How many probe vehicles do we need to collect on-street parking information? In: 2017 5th IEEE International Conference on Models and Technologies for Intelligent Transportation Systems (MT-ITS), pp. 538–543. IEEE (2017)
8. Bock, F., Liu, J., Sester, M.: Learning on-street parking maps from position information of parked vehicles. In: Sarjakoski, T., Santos, M.Y., Sarjakoski, L.T. (eds.) Geospatial Data in a Changing World. LNGC, pp. 297–314. Springer, Cham (2016). https://doi.org/10.1007/978-3-319-33783-8_17

9. Coric, V., Gruteser, M.: Crowdsensing maps of on-street parking spaces. In: 2013 IEEE International Conference on Distributed Computing in Sensor Systems, pp. 115–122. IEEE, May 2013. http://ieeexplore.ieee.org/xpls/abs_all.jsp?arnumber=6569416. http://ieeexplore.ieee.org/lpdocs/epic03/wrapper.htm?arnumber=6569416

10. SFMTA: SFpark: putting theory into practice. Pilot project summary and lessons learned (2014). http://sfpark.org/resources/docs_pilotsummary/. Accessed 24 June 2016

11. Kotb, A.O., Shen, Y., Huang, Y.: Smart parking guidance, monitoring and reservations: a review. IEEE Intell. Transp. Syst. Mag. 9(2), 6–16 (2017)

12. Ganti, R.K., Ye, F., Lei, H.: Mobile crowdsensing: current state and future challenges. IEEE Commun. Mag. 49(11), 32–39 (2011)

13. Rinne, M., Törmä, S., Kratinov, D.: Mobile crowdsensing of parking space using geofencing and activity recognition. In: 10th ITS European Congress, Finland, Helsinki, pp. 16–19 (2014)

14. Mathur, S., et al.: ParkNet: drive-by sensing of road-side parking statistics. In: Proceedings of the 8th International Conference on Mobile Systems, Applications, and Services, pp. 123–136. ACM, New York (2010)

15. Bock, F., Di Martino, S., Sester, M.: What are the potentialities of crowdsourcing for dynamic maps of on-street parking spaces? In: Proceedings of the 9th ACM SIGSPATIAL International Workshop on Computational Transportation Science, IWCTS 2016, pp. 19–24. ACM, New York (2016). http://doi.acm.org/10.1145/3003965.3003973

16. Richter, F., Di Martino, S., Mattfeld, D.C.: Temporal and spatial clustering for a parking prediction service. In: 2014 IEEE 26th International Conference on Tools with Artificial Intelligence (ICTAI), pp. 278–282. IEEE (2014)

17. Tasseron, G., Martens, K., van der Heijden, R.: The potential impact of vehicle-to-vehicle communication on on-street parking under heterogeneous conditions. IEEE Intell. Transp. Syst. Mag. 8(2), 33–42 (2016)

18. Waterson, B., Hounsell, N., Chatterjee, K.: Quantifying the potential savings in travel time resulting from parking guidance systems—a simulation case study. J. Oper. Res. Soc. 52(10), 1067–1077 (2001)

19. Benenson, I., Martens, K., Birfir, S.: PARKAGENT: an agent-based model of parking in the city. Comput. Environ. Urban Syst. 32(6), 431–439 (2008). http://linkinghub.elsevier.com/retrieve/pii/S0198971508000689

20. Millard-Ball, A., Weinberger, R.R., Hampshire, R.C.: Is the curb 80% full or 20% empty? Assessing the impacts of San Francisco's parking pricing experiment. Transp. Res. Part A: Policy Pract. 63, 76–92 (2014). http://linkinghub.elsevier.com/retrieve/pii/S0965856414000470

21. Bonsall, P., Palmer, I.: Modelling drivers' car parking behaviour using data from a travel choice simulator. Transp. Res. Part C: Emerg. Technol. 12(5), 321–347 (2004)

Continuous Nearest Neighbor Query
with the Direction Constraint

Xue Miao, Xi Guo, Huan Wang[✉], Zhaoshun Wang, and Xiaodong Ye

Beijing Key Laboratory of Knowledge Engineering for Materials,
University of Science and Technology Beijing, Beijing, China
xuemiao@xs.ustb.edu.cn, xiguo@ustb.edu.cn, wanghuan615@126.com,
zhswang@sohu.com, xdongye@126.com

Abstract. This paper proposes a new spatial data query, the continuous nearest neighbor query with the direction constraint (DCNN query). The DCNN query finds the nearest point of interest (POI), which also satisfies the direction constraint. The direction constraint depends on the direction of the user's speed. It is an angle range with the direction of the user's speed as the angle bisector. And the angle range is determined by the user. We study the snapshot and the continuous DCNN queries. The snapshot query finds the nearest POI according to the user's current location. The continuous query updates the results when the user is moving on a segment. To answer the DCNN queries, we propose R-tree-based approaches. The snapshot query algorithm is more efficient than the existing method. The continuous query algorithm finds the change points of the results according to two geographical properties. The experimental results show that we can answer the snapshot and the continuous queries accurately and efficiently.

Keywords: Continuous query · Location-based services ·
Direction constraint · Spatial databases

1 Introduction

Spatial data queries are widely used in geographic information systems (GIS), computer-aided designs (CAD), multimedia systems and so on. In recent years, with the rapid developments of the electronic intelligent equipments, location-based services (LBS) are growing rapidly in mobile social networks. Typically, the LBS recommend POIs according to the users' locations [1,2]. On the one hand, considering the spatial proximity of spatial objects, it is a good choice to recommend the nearest neighbor to the user [3]. On the other hand, directions are also important as well as distances. In order to support direction-based spatial data query, we study the continuous nearest neighbor query with the direction constraint (DCNN query).

    The DCNN query retrieves the nearest POI that is in the acceptable angle range. As Fig. 1 shows, the direction of the user's speed is q_{md}. The angle between

© Springer Nature Switzerland AG 2019
Y. Kawai et al. (Eds.): W2GIS 2019, LNCS 11474, pp. 85–101, 2019.
https://doi.org/10.1007/978-3-030-17246-6_8

q_{md} and the positive x-axis is ω, which denotes the user's moving direction. The user's acceptable angle range depends on a small angle θ, which is shown as the deflection on the left or right side of q_{md}. Figure 1(a) shows a snapshot DCNN query. The current location of the user is q. The POI p_1 is the nearest one to the user in the acceptable angle range. The snapshot query recommends p_1 to the user. Figure 1(b) and (c) show a continuous DCNN query. As Fig. 1(b) shows, when the user is at q_x, $p_1, p_2, \ldots, p_7$ are in the user's acceptable angle range, and p_1 is the nearest one. We recommend p_1 to the user. As Fig. 1(c) shows, when the user moves from q_x to q_y, p_2, p_3, and p_6 are in the acceptable angle range. The POI p_1 is no longer in the acceptable angle range and p_2 is the nearest one. The user moves on segment se, where s is his start point and e is his end point, the continuous query can dynamically update the results.

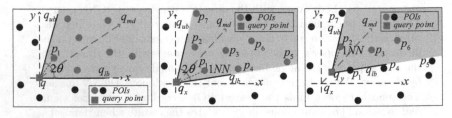

(a) Snapshot DCNN query   (b) Query point is at q_x (c) Query point is at q_y

Fig. 1. Examples of the DCNN query

Application Scenario. In computer games, such as Counter-Strike:Global Offensive, a counter strike can use the DCNN query to find the nearest terrorist that is the nearest one in the shooting angle range.

The contributions of our paper are summarized as follows:

- We propose the continuous nearest neighbor query with the direction constraint, which the existing work [17] does not focus on.
- We propose an efficient algorithm to answer the continuous query, and also improve the snapshot query algorithm.
- Experimental results show that the proposed algorithms are efficient.

The rest of our paper is organized as follows. The related work is summarized in Sect. 2. The formal definition of the DCNN query is proposed in Sect. 3. The algorithm of the snapshot query is shown in Sect. 4. The pruning technology and two properties are proposed in Sects. 5.1 and 5.2, and the continuous query algorithm is shown in Sect. 5.3. The experimental results of the snapshot query and the continuous query are shown in Sect. 6.

2   Related Work

Nearest neighbor query [4] is a basic spatial data query problem. In recent years, the research on nearest neighbor problem has been extended to k nearest neighbor query [5], obstacle k nearest neighbor query [6,7], reverse nearest neighbor query [8–11], group nearest neighbor query [12], aggregate nearest neighbor query [13], visible nearest neighbor query [14] and so on. Direction-based spatial data query is a very valuable research subject in spatial data query. Some achievements have been made in recent years. Direction-based multiple views on data [15] proposes using the pie-shaped heat map to visualize the distribution of POIs around the user. When POIs are recommended for users, direction-based surrounder (DBS) query [3] takes into account both the directions and the distances of the POIs. And [3] also proposes a snapshot query algorithm and a continuous query algorithm. On the basis of [3], reverse direction-based surrounder (RDBS) query [16] is obtained, which focuses on the inverse relationships between the POIs and the users. Direction-aware spatial keyword search (DESKS) [17] finds the top-k nearest POIs that are in the angle range and contain the required keywords. Besides these, there is another direction-aware approach, i.e., towards why-not spatial keyword top-k queries [18].

Direction-Aware Spatial Keyword Search (DESKS). $\langle(q.x, q.y) ; [\alpha, \beta] ; \mathcal{K}; k\rangle$ denotes the query point q. The location of q is $(q.x, q.y)$, the $[\alpha, \beta]$ is a direction constraint, and the $\mathcal{K}$ is a set of user-input keywords. There is a set of POIs P, each POI $p_i \in P$ has a location $(p_i.x, p_i.y)$ and it also has a set of keywords, which is denoted by $p_i.d$. Therefore, let $\langle(p.x, p.y) ; p.d\rangle$ denote the POI p. Given a query point q, the DESKS finds a set of candidates. Each candidate in the set satisfies both the direction constraint and the keywords constraint. This means that all POIs in the set are in the angle range $[\alpha, \beta]$, and every candidate contains all the keywords in the set $\mathcal{K}$. The top-k nearest neighbors in the candidates set are the answers of the DESKS. In a word, the DESKS finds the top-k nearest POIs that are in the angle range and contain all keywords specified by the user. As far as we know, there are not many investigations on the direction-aware search. The DESKS can support direction-aware search and also meet the needs of the LBS users.

On the basis of the DESKS, we propose a new query, the continuous nearest neighbor query with the direction constraint (DCNN query). Different from the DESKS, the DCNN query focuses on not only the snapshot query, but also the continuous query. In addition, our algorithm is more efficient.

3   Preliminaries

In the two-dimensional Euclidean space, there is a set of POIs $P = \{p_1, p_2, \ldots, p_n\}$, and the user's location is q.

Definition 1 (Moving direction): The user's moving direction is the direction of his speed q_{md}. The direction is measured by the angle between q_{md} and the positive x-axis. Let ω denote the direction.

Definition 2 (Direction): The angle between the segment from q to the POI p_i and the positive x-axis is the direction of p_i. Let δ_i denote the direction of p_i. $\forall p_i \in P, \exists \delta_i, i \in \{1, 2, \ldots, n\}$.

Definition 3 (Distance): The euclidean distance of the segment from q to the POI p_i is the distance of p_i. Let $dist(q, p_i)$ denote the distance of p_i. $\forall p_i \in P, \exists dist(q, p_i), i \in \{1, 2, \ldots, n\}$.

Definition 4 (Acceptable angle range): The user's moving direction is ω, and the angles deviating from the left and the right sides of the user's speed direction are all θ. Therefore, an acceptable angle range of 2θ is obtained. And the angular bisector of the acceptable angle range is the user's speed direction. $\theta \in (0, \pi/4)$ is determined by the user.

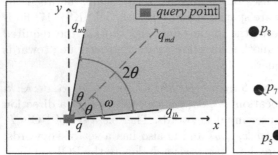
(a) Acceptable angle range

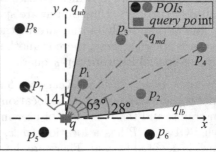
(b) Direction and distance

Fig. 2. Analyses of relevant definitions

For example, in Fig. 2, the user's current location is q. As Fig. 2(a) shows, the direction of the user's speed is q_{md}, and the angle between q_{md} and the positive x-axis is ω, so the user's moving direction is ω. In Fig. 2(a), there are two deflections on the left and right sides of the user's speed direction q_{md}. And the deflection θ is determined by the user, we call it acceptable angle. According to the moving direction ω and the acceptable angle θ, we can get an acceptable angle range of 2θ. As Fig. 2(b) shows, there is a set of POIs $P = \{p_1, p_2, p_3, p_4, p_5, p_6, p_7, p_8\}$. Some of the POIs are in the acceptable angle range, i.e., p_1, p_2, p_3, p_4, while some are not, i.e., p_5, p_6, p_7, p_8. In Fig. 2(b), for any POI $p_i \in P$ ($i \in \{1, 2, \ldots, 8\}$), we can make a segment from q to the POI p_i and the length of the segment qp_i is the distance of the POI p_i. There is an angle between the segment qp_i and the positive x-axis. The angle is the direction of the POI p_i. For example, the directions of the POIs p_1, p_4 and p_7 are 63°, 28° and 141°. Among the POIs within the acceptable angle range, p_1 is the nearest and p_4 is the farthest.

Definition 5 (DCNN): The DCNN is the nearest POI to the query point in the user's acceptable angle range.

Definition 6 (Snapshot DCNN query): The snapshot DCNN query finds the nearest POI $\in P$ to the query point in the user's acceptable angle range. Let $(q.x, q.y, \omega, \theta)$ denote the snapshot query, where the $(q.x, q.y)$ is the location of the user, the ω is the user's moving direction, and the θ is the user's acceptable angle.

As Fig. 1(a) shows, the query result is p_1, because p_1 is the nearest POI in the acceptable angle range.

Definition 7 (Continuous DCNN query): When the user moves on segment se, the continuous DCNN query dynamically finds the nearest POI in the user's acceptable angle range. Let $(s_x, s_y, e_x, e_y, \theta)$ denote the continuous query, where the (s_x, s_y) is the start point, the (e_x, e_y) is the end point of the segment se, and the θ is the user's acceptable angle.

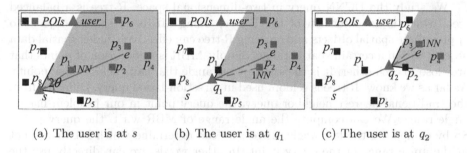

(a) The user is at s (b) The user is at q_1 (c) The user is at q_2

Fig. 3. An example of the continuous DCNN query

Figure 3 is an example of the continuous DCNN query. There is a POIs set $P = \{p_1, p_2, p_3, p_4, p_5, p_6, p_7, p_8\}$. The user moves on segment se from s to e. As Fig. 3(a) shows, the user's current location is s and POIs p_5, p_7, p_8 are not in the acceptable angle range of 2θ and p_1, p_2, p_3, p_4, p_6 are in it. Because p_1 is the nearest one in the acceptable angle range, we recommend it to the user. When

Table 1. Symbols and descriptions.

Symbol	Description
q	Query point (The user)
ω	The user's moving direction
θ	The user's acceptable angle
P	A collection of data objects in two-dimensional Euclidean space
δ_i	The direction of POI p_i, $p_i \in P$
$dist(q, p_j)$	The distance of POI p_j, $p_j \in P$
se	se is a line segment, the user moves on it
q_{md}	Direction of user's speed, the angle between it and the positive x-axis is ω

the user moves to q_1, p_1 is not in the acceptable angle range. We recommend p_2 to the user, because it is the nearest one in the acceptable angle range. In Fig. 3(c), when the user moves to q_2, only p_3 and p_4 are in the acceptable angle range, and we recommend the nearest one p_3. Table 1 summarizes the symbols used in this paper.

4   Snapshot DCNN Query

In this section, we introduce how to answer the snapshot query. The challenge is to find the objects in the acceptable angle range efficiently. We calculate the angle ranges of MBRs relative to the query point, and determine which MBRs are not in the acceptable angle range by comparing the angle ranges of MBRs and the acceptable angle range. The spatial objects that are in the MBRs, which are outside the acceptable angle range, should be filtered out.

We study the DCNN query in two-dimensional space. R-tree is a balanced tree, which is an extension of the B-tree. Using MBRs of spatial objects to approximate spatial objects and creating R tree can effectively index spatial data objects within a certain spatial range. Simply, MBR is a collection of points that are close to each other. It is the smallest rectangle that can enclose these points. As far as we know, R tree is seldom used in direction-based query. Different from the traditional nearest neighbor query, the query point in our problem has an angle range. We can compute the angle range of MBR w.r.t. the query point. So by comparing the two angle ranges, we can determine which MBRs are not in the angle range of the query point. In other words, we can directly use the user's angle range to prune the MBRs and speed up the query. Therefore, we use R tree as the index structure.

In the snapshot query algorithm, we use a priority queue. The initial value of the priority queue is a tuple, which consists of the distance (considered to be 0) from the root node to the query point and the root node. The priority is the distance from the node to the query point. The shorter the distance from the node to the query point, the higher the priority is. Each time the node with the highest priority is popped. If the popped node is an internal node (branch node), it's child nodes should be checked. If the child node is within or intersects with the acceptable angle range, it should be pushed into the queue with the priority, which is the distance from the child node to the query point. If the node popped is a leaf node, its interior points should be checked. If the point is in the acceptable angle range, it should be pushed into the queue with the distance from it to the query point. If the popped node is a point, it is a result. And the counter i plus one (the initial value of i is 0). When $i < k$, we will repeat the above operations. When $i = k$, we find all results and end the program.

Algorithm 1. Get$k$DCNNs(que, $query$)

Input: $q.x$, $q.y$, ω, θ, k

Output: the top-k DCNNs of the query point q

1    $node \leftarrow$ pop the first element of the priority queue que.

2   **if** $node$ *is* Rtree.Branch **then**

3   **for** *child* in *node*.childList **do**

4   **if** *child* is in or intersects with the acceptable angle range **then**

5             $que \leftarrow ((dist(child,query),child),\ dist(child,query))$

6   **if** $node$ *is* Rtree.Leaf **then**

7   **for** *point* in *node*.childList **do**

8   **if** *point* is in the acceptable angle range **then**

9             $que \leftarrow ((dist(point,query),point),\ dist(point,query))$

10   **if** $node$ *is* Rtree.Point **then**

11       $i{+}{=}1$

12   **if** $i{==}k$ **then**

13   show the results and end the program

Algorithm 1: The algorithm finds the top-k nearest POIs. The coordinate $(q.x, q.y)$ is the location of the user. The user's moving direction is ω, and his acceptable angle is θ. From line 2 to line 5, we first determine the type of the node popped. If it is an internal node, we determine whether its child node *child* is in or intersects with the acceptable angle range (Line 4). If it is, then we calculate the distance $dist(child, query)$ from the child node *child* to the query point *query* and push the child node with the distance into the priority queue *que* (Line 5). From line 6 to 9, if the node popped is a leaf, we determine whether the child node is in the acceptable angle range (Line 8). If it is, we execute line 9. From line 10 to 13, if the node popped is a point, it is a result. When $i == k$, we show the results and terminate the program.

5   Continuous DCNN Query

In this section, we introduce how to answer the continuous query. The challenge is to find the change points accurately. In our algorithm, the change points are determined one by one. We first calculate the vertical bisector between the first DCNN and the second DCNN. Then we calculate the intersection point between the vertical bisector and the segment *se*. We judge whether the possible change point is accurate according to the position of the intersection point.

5.1   Pruning Rule

In order to answer the continuous query more effectively, we first discuss how to prune the MBRs. In addition, we use priority queue to find the results as soon as possible and to end the search. We use the pruning rule to handle the nodes of R-tree to get the query results we want. In order to get our pruning rule, we should first solve the following problems.

– How to calculate the acceptable angle range of the user?
– How to calculate the angle range of MBR w.r.t. the user?

According to the sizes of ω and θ, the acceptable angle range includes three cases. **Case 1**, $\omega \geq \theta$ and $\omega + \theta \leq 2\pi$ (Fig. 4(a)). **Case 2**, $\omega > \theta$ and $\omega + \theta > 2\pi$ (Fig. 4(b)). **Case 3**, $\omega < \theta$ (Fig. 4(c)).

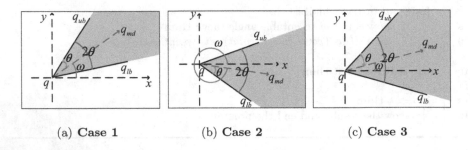

(a) **Case 1**   (b) **Case 2**   (c) **Case 3**

Fig. 4. Three cases of acceptable angle range

– **Case 1:** Because $\theta \leq \omega$ and $\omega + \theta \leq 2\pi$, the angles of q_{ub} and q_{lb} are as the following formulas

$$q_{ub} = \omega + \theta \tag{1}$$

$$q_{lb} = \omega - \theta \tag{2}$$

– **Case 2:** Because $\theta + \omega > 2\pi$, the angle of q_{ub} cannot be determined directly by $\omega + \theta$. So we need to use the following formulas

$$q_{ub} = \omega + \theta - 2\pi \tag{3}$$

$$q_{lb} = \omega - \theta \tag{4}$$

– **Case 3:** Because $\theta > \omega$, the angle of q_{lb} cannot be determined directly by $\omega - \theta$. So we need to use the following formulas

$$q_{ub} = \omega + \theta \tag{5}$$

$$q_{lb} = 2\pi + \omega - \theta \tag{6}$$

There are three kinds of location relationships between the acceptable angle range and the nodes of R-tree, which are: (1) MBR in the acceptable angle range (For example, the mbr_1 in Fig. 5(a)); (2) MBR intersects with the acceptable angle range (For example, the mbr_2 in Fig. 5(b)); (3) MBR is outside the acceptable angle range (For example, the mbr_3 in Fig. 5(c)).

Pruning Rule: If some nodes of R-tree are outside the acceptable angle range, that is to say, these nodes all belong to (3) position relationship. It is impossible for spatial data objects that are in such nodes to be in the acceptable angle range. Therefore, we can prune such nodes.

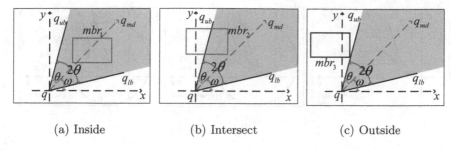

(a) Inside (b) Intersect (c) Outside

Fig. 5. The location relationship between acceptable angle range and MBR

5.2 Properties

Property 1: The user moves on the line segment se. When the user is located at q_i of se, p_{i+1} is an object within the user's acceptable angle range. The projection of p_{i+1} parallel to the upper or lower boundary line of user's acceptable angle range on se is q_{i+1}. When the user moves from q_i to q_{i+1}, p_{i+1} is always within the acceptable angle range, but when the user arrives at or leaves q_{i+1}, p_{i+1} will no longer be within the acceptable angle range. If p_{i+1} is the user's DCNN at q_i, q_{i+1} may be a change point.

Proof. In Fig. 6(a), se is a line segment, the user's speed direction is from s to e, and the angle between it and the positive x-axis is ω. The user's current position is q_i, and p_{i+1} is a POI that is in the user's acceptable angle range. q_{ub} (q_{lb}) is the upper (lower) boundary line of the user's acceptable angle range. q_{i+1} is the projection of p_{i+1} parallel to q_{ub} on se. From q_i to q_{i+1}, p_{i+1} is always in the user's acceptable angle range. But when the user arrives at or leaves q_{i+1}, p_{i+1} will not appear in the acceptable angle range. Assuming that the user has reached the position q_{i+j}, it is obvious that p_{i+1} is not in the acceptable angle range. As Fig. 6(b) shows, let q_i be the center and the distance from q_i to p_{i+1} be the radius of the circle q_i. The intersection between q_{ub} (q_{lb}) and the circle q_i is u (l). If p_{i+1} is the DCNN of the user at q_i, there are no other points in the sector uq_il, which can be denoted by S_{uq_il}. As Fig. 6(c) shows, let q_{i+1} be the center and the distance from q_{i+1} to p_{i+1} be the radius of the circle q_{i+1}. There are no points in the intersection of S_{uq_il} and circle q_{i+1}. We suppose the user's DCNN at q_{i+1} is p_{i+2}. If p_{i+2} is in circle q_{i+1} but not in circle q_i, p_{i+2} will replace p_{i+1} as the user's DCNN when the user is at somewhere of segment q_iq_{i+1}, although p_{i+1} is still in the acceptable angle range. In this case, q_{i+1} isn't a real change point. However, if p_{i+2} is not in the circle q_{i+1}, p_{i+1} is always the DCNN when the user moves from q_i to q_{i+1}. When the user arrives at or leaves q_{i+1}, the user's DCNN is p_{i+2} because p_{i+1} isn't in the acceptable angle range. In this case, q_{i+1} is a real change point. If q_{i+1} isn't a real change point, we need to locate the real change point through p_{i+1} and p_{i+2}. That is Property 2.

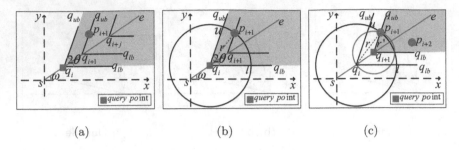

(a)   (b)   (c)

Fig. 6. Proof of Property 1

Property 2: In the continuous DCNN query, the user moves on the segment se. The current query results are $[\ldots, (q_{i-1}, p_i), (q_i, p_{i+1})]$. The q_{i-1} is a real change point and q_i is a possible change point. The POIs p_i and p_{i+1} are DCNNs. The sufficient and necessary condition of q_i to be a real change point is $dist(q_{i-1}, q_i) \leq dist(q_{i-1}, q_i')$. The q_i' is the intersection point between the straight line l_{vb} and the segment se. The straight line l_{vb} is the vertical bisector between p_i and p_{i+1}. $dist(q_{i-1}, q_i)$ is the distance from q_{i-1} to q_i.

Proof. In Fig. 7, the user's acceptable angle is θ and the user's moving direction is ω. The current query results are $[(s, p_1), (q_1, p_2)]$, where s is a real change point, and q_1 is a possible change point. The POIs p_1 and p_2 are their DCNNs. The straight line l_{vb} is the vertical bisector between p_1 and p_2. The point q_1' is the intersection point between the line l_{vb} and the segment se. As Fig. 7(a) shows, $dist(s, q_1') < dist(s, q_1)$. From s to q_1', the user's DCNN is p_1 and from q_1' to q_2 the user's DCNN is p_2. So q_1 isn't a real change point and q_1' is the real change point. As Fig. 7(b) and (c) show, $dist(s, q_1') > dist(s, q_1)$. When the user moves from s to q_1, the user's DCNN is p_1. And when the user arrives at or leaves q_1, the DCNN is p_2. So q_1 is a real change point.

We cannot directly determine that the possible change point q_1 is a real change point. There is a case that when the user is located somewhere on the segment sq_1, although p_1 is within the acceptable angle range, p_2 becomes the DCNN of the user instead of p_1. So only p_1 and p_2 need to be considered in

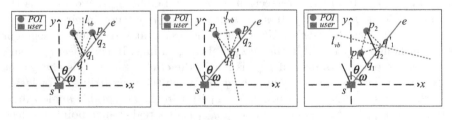

(a) $dist(s, q_1') < dist(s, q_1)$ (b) $dist(s, q_1') \geq dist(s, q_1)$ (c) $dist(s, q_1') \geq dist(s, q_1)$

Fig. 7. Proof of Property 2

determining where the real change point is. Therefore, only two adjacent DCNNs need to be considered in the subsequent determination of the real change point.

5.3   Query Algorithm

We propose an algorithm (Algorithm 2) to answer the continuous DCNN query. The continuous query algorithm uses the Property 1 and the Property 2. On the basis of the snapshot algorithm, if the node popped from the priority queue is a point, operations are as follows.

Algorithm 2: The user moves on the segment se and the coordinates (s_x, s_y) and (e_x, e_y) are the start point and the end point of se. We can calculate the moving direction according to the segment se. The angle θ is the user's acceptable angle. Line 2, we use a list ret to store the change point and it's DCNN. Line 3, we use l to count the number of the elements in ret. Line 4, we calculate the projection of the current DCNN $node$ parallel to the upper or lower boundary line of the acceptable angle range on se. The upper or lower boundary line of the acceptable angle range is q_{ub} or q_{lb}. Line 6, when $l \geq 2$, we compute the vertical bisector (l_{vb}) between two adjacent DCNNs $ret[l-2][1]$ and $ret[l-1][1]$. Line 7, we calculate the intersection point ($inpoint$) between the vertical bisector l_{vb} and the segment se. If the condition of line 8 holds, i.e., the distance from the change point $ret[l-2][1]$ to the intersection point $inpoint$ is less than the distance from the change point to the possible change point $ret[l-1][0]$. In this case, the possible change point $ret[l-1][0]$ is not a real change point and the $inpoint$ is a real change point. Then we use the real change point $inpoint$ to replace the possible change point $ret[l-1][0]$ (Line 9). Line 10, we judge whether to terminate the program, otherwise we empty and initialize the priority queue que and continue the above operations.

Algorithm 2. GetAcDCNN(que, $query$)

Input: s_x, s_y, e_x, e_y, θ
Output: the change points and theirs DCNNs

```
1   if node is Rtree.Point then
2       ret←((query,node))
3       l+=1
4       query←get the projection of node parallel to q_ub or q_lb on se
5       if l>1 then
6           l_vb ←get the vertical bisector of ret[l-2][1] and ret[l-1][1]
7           inpoint←get the intersection between l_vb and se
8           if dist(ret[l-2][0], inpoint)< dist(ret[l-2][0], retx[l-1][0]) then
9               ret[l-1][0]←inpoint
10              if dist(s,e)< dist(s,ret[l-1][0]) then
11                  print ret and end the program
12              else
13                  empty and initialize the priority queue que
14      else
15          empty and initialize the priority queue que
```

6   Experiments

This section reports the performances of the proposed algorithms. All algorithms are implemented in Python and all experiments are carried out on the ubuntu 16.04 system.

Table 2. Synthetic and real datasets.

Dataset size (synthetic and real)	Region size (synthetic and real)	Data type (real)
0.1k	$[0, 100] \times [0, 100]$	Hotels
1k	$[0, 100] \times [0, 1000]$	Gas stations
10k	$[0, 1000] \times [0, 1000]$	Banks
100k	$[0, 1000] \times [0, 10000]$	Restaurants

Table 2 shows the datasets we use in this paper, which include the real datasets and the synthetic datasets. The synthetic datasets are generated randomly. The real datasets are crawled from the High German map. They are the data of hotels, gas stations, banks and restaurants in Beijing, China.

6.1   Experiments on Snapshot Queries

In this section, we show the experimental results of the snapshot algorithm and the existing algorithm. Let "DCNN" and "DESKS" denote the snapshot and the existing algorithms. We explore the performances of the algorithms by varying the acceptable angle sizes, the dataset sizes and the number of points in MBR. The vertical axis represents the query time (millisecond) and the horizontal axis represents the influence factors.

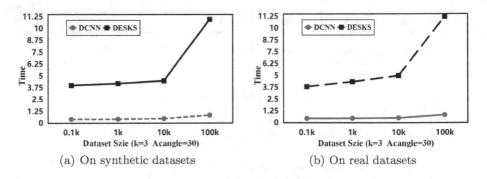

(a) On synthetic datasets   (b) On real datasets

Fig. 8. Experiments of varying the dataset sizes

As Fig. 8(a) and (b) show, they are the experimental results of varying the dataset sizes. When the dataset size is less than or equal to 10000, our algorithm is better than the existing algorithm, but the effect is not obvious. We can find that our algorithm is more efficient than the existing algorithm when the dataset size is larger than 10000. Through the above analyses, we find that the existing method may not be suitable for the larger datasets. "Acangle" is the acceptable angle and the number of points in MBR is 25.

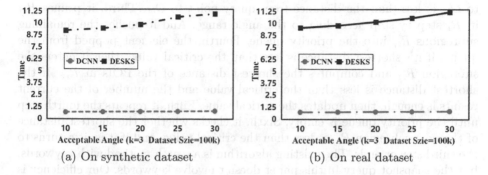

(a) On synthetic dataset   (b) On real dataset

Fig. 9. Experiments of varying the acceptable angle sizes

In Fig. 9(a) and (b) are the experimental results of varying the acceptable angle sizes. We can find that the snapshot algorithm is more efficient than the existing algorithm when the dataset size is 100000 and the acceptable angle (i.e., the acceptable angle range) sizes are different. According to the user's moving direction ω and the user's acceptable angle θ, we can calculate the user's acceptable angle range. The larger the acceptable angle, the larger the acceptable angle range is. The number of the points in MBR is 25.

Figure 10(a) and (b) are the experimental results of our snapshot algorithm of varying the number of points in MBR. When the dataset size is 100, and the number of points in MBR is 10, the snapshot algorithm has the highest efficiency. And when the dataset size is 1000 or 10000, and the number of points in MBR is 15, or when the dataset size is 100000, and the number of points in MBR is 25, the snapshot algorithm works best.

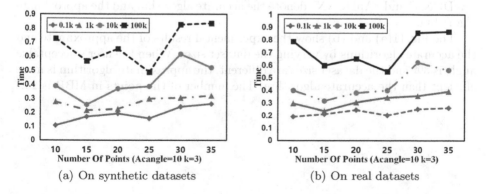

(a) On synthetic datasets   (b) On real datasets

Fig. 10. Experiments of varying the number of points in MBR

The existing algorithm first computes the minimum boundary rectangle of all objects in space. These POIs are divided into different regions R_i according to their distances to the bottom left of the MBR (O_{bl}). For POIs in each region R_i, they are divided into different subregions R_{ij} according to their directions relative to O_{bl} of the MBR. Second, it prunes the region R_i whose distance to O_{bl} is less than the distance from query point to O_{bl}. Third, it prunes R_{ij} in R_i step by step according to the angle range. And it pushes the remaining subregions R_{ij} into the priority queue. Fourth, the element popped from the queue, if it's shortest distance is less than the critical value, then it enters the subregion R_{ij} and computes the shortest distance of the POIs in R_{ij}. If the shortest distance is less than the critical value and the number of the current results is enough, then updates the critical value. Fifth, it repeats the fourth step until the priority queue is empty. Sixth, it checks whether the shortest distance of the next subregion R_{ij} is less than the critical value, and if it is, it returns to the third step or ends. The existing algorithm is a search method with keywords, but the snapshot query in this paper doesn't involve keywords. Our efficiency is higher, it is very likely that the existing algorithm is not suitable for searching without keywords. In addition, a lot of distance calculations are used in the query process. Another possibility is that our datasets are not suitable for the existing algorithm.

6.2   Experiments on Continuous Queries

In this section, we evaluate the effect by varying the dataset sizes, the acceptable angle sizes, and the number of points in MBR. In addition, we show the accuracy of the approximate algorithm. The unit of the query time is millisecond. We propose two algorithms to answer the continuous query, the accurate algorithm and the approximate algorithm. The approximate algorithm only uses the Property 1, and the change points computed by this algorithm may not be accurate. Therefore, we propose an accurate algorithm, which uses the Property 1 and the Property 2. The accurate algorithm can exactly find all change points and their DCNNs. We compare the efficiencies of the two algorithms. Let "AcDCNN" and "ApDCNN" denote the accurate algorithm and the approximate algorithm.

The Fig. 11(a) and (b) show the experimental results of the approximate and the accurate algorithms by varying the dataset sizes. When the user's acceptable angle is 30° and the dataset sizes are different, the approximate algorithm is more efficient than the accurate algorithm. The number of the points in MBR is 25.

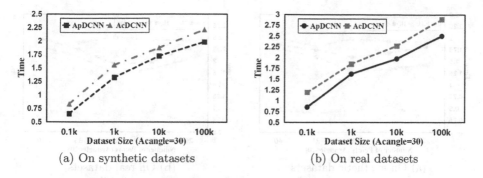

Fig. 11. Experimental results of varying the dataset sizes

The Fig. 12(a) and (b) show the experimental results of the two algorithms by varying the acceptable angle sizes. When the dataset size is 100000 and the acceptable angle sizes are different, the approximate algorithm is more efficient than the accurate algorithm. The number of the points in MBR is 25.

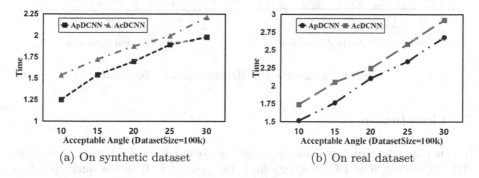

Fig. 12. Experimental results of varying the acceptable angle sizes

As Fig. 13(a) and (b) show, when the number of points in MBR is different, the efficiency of the accurate algorithm is different. When the dataset size is 100 and the number of points in MBR is 20, the accurate algorithm has the highest efficiency. When the dataset size is 1000 or 10000 and the number of points in MBR is 30, or when the dataset size is 100000 and the number of points in MBR is 35, the accurate algorithm works best.

Figure 14(a) and (b) are the experimental results of the accuracy of the approximate algorithm. When the dataset size is the same and the acceptable angle size is different, the accuracy of the approximate algorithm is different. And when the acceptable angle size is the same, and the dataset size is different, the accuracy is different. The accuracy is obtained by comparing the experimental results of the accurate and the approximate algorithms. The number of the points in MBR is 40.

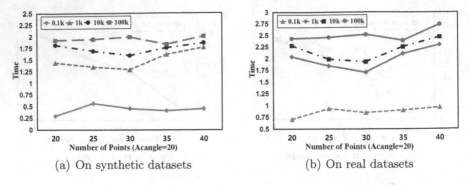

(a) On synthetic datasets   (b) On real datasets

Fig. 13. Experimental results of varying the number of points in MBR

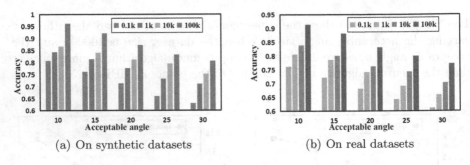

(a) On synthetic datasets   (b) On real datasets

Fig. 14. The accuracy of the approximate algorithm

7   Conclusion

In this paper, we study the nearest neighbor query with the direction constraint (DCNN query). The DCNN query finds the nearest POI to the query point in the user's acceptable angle range. We propose algorithms to answer the snapshot and the continuous queries. In snapshot query, our algorithm is better than the existing algorithm. Our continuous query algorithm can find the change points and update the results efficiently.

Acknowledgment. This work is supported by the National Natural Science Foundation of China (No. 61602031). This work is also supported by Fundamental Research Funds for the Central Universities (No. FRF-TP-16- 011A3, No. FRF-BD-16-010A). In addition, we appreciate the valuable comments from Prof. Dezheng Zhang.

References

1. Schlegel, R., Chow, C.Y., Huang, Q., Wong, D.S.: Privacy-preserving location sharing services for social networks. IEEE Trans. Serv. Comput. **10**(5), 811–825 (2017)
2. Li, L., Lu, R., Huang, C.: EPLQ: efficient privacy-preserving location-based query over outsourced encrypted data. IEEE Internet Things J. **3**(2), 206–218 (2016)

3. Guo, X., Zheng, B., Ishikawa, Y., Gao, Y.: Direction-based surrounder queries for mobile recommendations. VLDB J. **20**(5), 743–766 (2011)
4. Roussopoulos, N., Kelley, S., Vincent, F.: Nearest neighbor queries. In: Proceedings of the ACM International Conference on Management of Data (SIGMOD), vol. 24, no. 2, pp. 71–79 (1995)
5. Kim, H., Chang, J.: K-nearest neighbor query processing algorithms for a query region in road networks. J. Comput. Sci. Technol. **28**(4), 585–596 (2013)
6. Gu, Y., Yu, G., Yu, X.: An efficient method for k nearest neighbor searching in obstructed spatial databases. J. Inf. Sci. Eng. **30**(5), 1569–1583 (2014)
7. Yu, X.N., Gu, Y., Zhang, T.C., Yu, G.: A method for reverse k-nearest-neighbor queries in obstructed spaces. Chin. J. Comput. **34**(10), 1917–1925 (2011)
8. Safar, M., Ibrahimi, D., Taniar, D.: Voronoi-based reverse nearest neighbor query processing on spatial networks. Multimedia Syst. **15**(5), 295–308 (2009)
9. Achtert, E., Bhm, C., Krger, P., Kunath, P., Pryakhin, A., Renz, M.: Efficient reverse k-nearest neighbor search in arbitrary metric spaces. In: SIGMOD Conference, pp. 515-526 (2006)
10. Tao, Y., Yiu, M., Mamoulis, N.: Reverse nearest neighbor search in metric spaces. IEEE Educ. Activities Dept. **18**(9), 1239–1252 (2006)
11. Benetis, R., Jensen, C., Karciauskas, G., Saltenis, S.: Nearest and reverse nearest neighbor queries for moving objects. VLDB J. **15**(3), 229–249 (2006)
12. Papadias, D., Shen, Q., Tao, Y., Mouratidis, K.: Group nearest neighbor queries. In: Proceedings of the 20th IEEE International Conference on Data Engineering, pp. 301–312 (2004)
13. Papadias, D., Tao, Y., Mouratidis, K., Hui, C.: Aggregate nearest neighbor queries in spatial databases. ACM Trans. Database Syst. (TODS) **30**(2), 529–576 (2005)
14. Nutanong, S., Tanin, E., Zhang, R.: Incremental evaluation of visible nearest neighbor queries. IEEE Trans. Knowl. Data Eng. **22**(5), 665–681 (2010)
15. Guo, X., Yu, J., Li, R., Miao, X.: Direction-based multiple views on data. World Wide Web **22**, 1–35 (2018)
16. Guo, X., Ishikawa, Y., Wulamu, A., Xie, Y.: Reverse direction-based surrounder queries. In: Cheng, R., Cui, B., Zhang, Z., Cai, R., Xu, J. (eds.) APWeb 2015. LNCS, vol. 9313, pp. 280–291. Springer, Cham (2015). https://doi.org/10.1007/978-3-319-25255-1_23
17. Li, G., Feng, J., Xu, J.: DESKS: direction-aware spatial keyword search. In: IEEE International Conference on Data Engineering, vol. 1084, no. 4627, pp. 474–485 (2012)
18. Chen, L., Li, Y., Xu, J., Jensen, C.S.: Towards why-not spatial keyword top-k queries: a direction-aware approach. IEEE Trans. Knowl. Data Eng. **30**(4), 796–809 (2017)

Predicting the Spatial Impact of Planned Special Events

Sergio Di Martino[1(✉)], Simon Kwoczek[2], and Silvia Rossi[1]

[1] Department of Electrical and Telecommunications Engineering,
University of Naples Federico II, Naples, Italy
{sergio.dimartino,silvia.rossi}@unina.it
[2] L3S Research Center, Hannover, Germany
skwoczek@l3s.de

Abstract. While it is widely acknowledged that Planned Special Events (PSEs), such as concerts, soccer games, etc., have a strong impact on road traffic, very few studies have quantitatively investigated this phenomenon. In this paper we present the preliminary results of a technique to quantify the impact of PSEs on traffic around the venue of the events. In particular, our goal was to automatically identify all those road segments around a venue that show a different traffic behavior on event days than on non-event days. To this aim, we defined a specific pipeline, including a K-Nearest Neighbor classifier, trained on traffic data of event and non-event days for each road, using the Dynamic Time Warp (DTW) as distance metric. The proposed solutions has been empirically evaluated on four PSE venues in Germany. Two of them hosted only soccer matches of the German First League, while the other two had mixed types of PSEs, including sport, concerts and other categories of events. Results are very positive for the soccer stadiums, while more research is needed for the venues hosting mixed types of PSEs.

1 Introduction and Related Work

In times of steady growth of mobility demands, traffic congestion cost billions of dollars to the society every year [1]. This problem can be alleviated by the use of IT solutions enabling a smarter use of the road infrastructure. As an example, more reliable traffic predictions coming from the analysis of massive mobility dataset make it possible to reroute drivers towards less congested street segments at a given time, balancing the demand on the road network and allowing drivers to reach their destinations faster.

In order to obtain such predictions, in the last years many investigations have been conducted on the impact of different influencing factors, as accidents, construction zones, etc., on traffic (e.g. [2–4]). In particular, several data-driven approaches have been presented in the literature, including generalized linear regression [5], nonlinear time series [6], Support Vector Machines [7], Kalman filters [8], and various neural network models [8–10]. A comprehensive survey on these statistical approaches can be found in [11]. At the same time, thanks

© Springer Nature Switzerland AG 2019
Y. Kawai et al. (Eds.): W2GIS 2019, LNCS 11474, pp. 102–117, 2019.
https://doi.org/10.1007/978-3-030-17246-6_9

to the advances in sensor technologies (like Smartphones, GPS handhelds, etc.), massive datasets about traffic with high spatial and temporal resolution became available. The consequence is that today we have on the market services like TomTom MyDrive[1], or Google Maps[2], able to effectively report and predict the traffic situation in urban scenarios on mobile devices, such as smartphones or in-car navigation systems. The most of these services work well with recurring traffic situations, since they learn from the past what is the typical trend of the traffic on a given road segment at a give time/day of the week, and thus predict the future situation looking at this database of past mobility situations [12]. On the other hand, they fail in presence of situations that change the regular day-to-day pattern. Among them, there are the so called *Planned Special Events* (PSEs) [13], such as concerts, soccer games, etc., with large visitor numbers [2]. According to a study from the U.S. Federal Highway Administration [14], in the United States there are approximately 24,000 PSEs with more than 10,000 attendees, every year, and the nationwide congestion cost due to PSEs is between 1.7 and 3.4 billion US dollars per year. These numbers confirm the need to investigate PSE-specific influences, and to develop solutions that can be incorporated into future systems. In spite of these numbers, to date a very few investigations have been conducted on the impact of PSE on urban traffic [15–18].

Identifying the spatial impact of an influencing factor on traffic is a common issue for all the above mentioned works on mobility. However, PSEs bring in two additional problems: location and variability. As for the first, PSEs usually happen in dense inner city areas with complex traffic networks, where there is the problem of identifying the influenced roads within an intricate network topology, composed of thousands of roads and intersections. As for PSEs variability, as shown in [17], their impact changes a lot among different PSEs, due to many reasons, such as popularity of an event, target group of people and choice of transportation, etc. That, in combination with the absence of centralized information sources for PSEs, leads to difficulties in traffic strategy planning [19].

In [17] we conducted a first analysis on the impact of PSEs on traffic in inner city scenarios. In particular, we started from the observation that PSEs cause two subsequent waves of congestion [15], the first one due to people going to the event and the second one due to people leaving the venue. We investigated 29 events in the LANXESS arena in Cologne (Germany), using a dataset of about 50 GB of Floating Car Data collected around the venue. We defined a system based on Machine-Learning techniques to predict the impact of the second wave of traffic, given the first wave and the category of event (concert, sport, etc.). Results shown an improvement of the traffic predictions of more than 35%. In [18] we conducted a first analysis on the possibility to identify in a automatic way the road segments around a venue showing an anomalous traffic pattern in presence of PSEs, by exploiting Artificial Neural Networks. Preliminary results on two soccer stadiums in Germany were encouraging.

---

[1] https://mydrive.tomtom.com.
[2] https://www.google.com/maps/.

In the present paper we extend our previous investigations, by varying the two key issues of PSEs, i.e. location and variability. Indeed, we investigate venues embedded in complex network topologies, hosting different categories of events, like concerts, soccer games, hockey games, etc. Since, as reported in literature [17], different PSE categories led to very different impacts on traffic, this new domain required the investigation of more robust machine-learning solutions. After some benchmarking, we defined a specific processing pipeline based on a K-Nearest Neighbour classifier, trained on traffic data of event and non-event days for each road, using the Dynamic Time Warp (DTW) as distance metric. To evaluate the proposal, we analysed the approach on four different venues in Germany. Two of them hosted only soccer games, the other two had mixed types of PSEs. The investigated impact areas cover more than 2000 road segments. The work presented in this paper is partly based on the PhD thesis of Kwoczek [20].

Summarizing, the main contributions of this paper are:

1. We propose an automatic approach to identify the road segments whose traffic patterns are affected by PSEs
2. We evaluate our approach on all the soccer games taking place in two of the biggest stadiums in Germany
3. We evaluate our approach on all the events taking place in two further venues in Germany, comparing results with the stadiums hosting only soccer games.

The remainder of this paper is structured as follows: In Sect. 2 we describe the problem at the hand and present our proposed approach. In Sect. 3 we describe the experimental setting we defined for the empirical assessment of the proposal, with a particular emphasis on the used datasets. In Sect. 4 we present the results of our investigations, while in Sect. 5 some final remarks and future research directions are pointed out.

2   The Proposed Approach

In this section we provide a description of the problem followed by a presentation of the proposed approach.

2.1   Preliminary Definitions

There are many different definitions of the concept of *event* in the traffic literature. One commonly used definition, from the Traffic Engineering Handbook [21], uses three major categories: *planned, unplanned,* and *emergency events,* where an event can belong to different categories at the same time (e.g., unplanned and emergency). In the rest of the paper, we will focus only on planned events.

Generally, traffic congestion can be divided into *recurrent congestion (RC)* and *non-recurrent congestion (NRC)* [22,23]. RC are mainly due to some factors that periodically influence the network infrastructure, such as daily commute traffic or weekend travellers [23]. Due to this recurring nature, commuters and traffic management centres usually know the time, location, extent and duration of RC events. Thus, even if RC is stressful, it is somehow predictable and manageable. NRC, on the other hand, is defined as an unusual congestion, whose time, location and duration are mostly unknown in advance. This type of disruptions are caused by external factors, like accidents, construction zones, PSE, etc., whose impact on the traffic network highly depends on local and transient conditions.

Different approaches have been proposed in the literature to discriminate whenever a traffic situation is a RC or it exceeds the habitual patterns and turns into NRC (e.g. [23]). This is a very complex task, which, to the best of our knowledge, has never been applied to PSEs.

2.2 Problem Description

Our goal is to define a technique to automatically identify the set of road segments that will be part of routes that people usually will use to get to a venue before a PSE starts. We define a road segment as being **affected** by a PSE, if it shows a Non-Recurrent Congestion on event days, in a way that must be detectable by an automatic classifier [20]. This definition intentionally excludes road segments that are always congested (e.g. bottlenecks in the infrastructure), since, even if they will most likely show congestions also in presence of PSEs, the observed traffic behavior can not be explained by the presence of PSEs solely. In general, those bottlenecks are already well captured by traditional learning systems, based on historic data.

In this work, we will not consider different traffic trends for different weekdays, as we assume that the routes people tend to take to attend a PSE are independent of that. A further investigation on this point can be an interesting evolution of this research.

In this paper we differentiate between two different sets of roads segments:

- *Inbound segments:* the set of streets that drivers mostly use to reach the venue, and
- *Outbound segments:* the set of streets that drivers mostly use to leave the venue, once the PSE is over.

Our approach can work similarly for Inbound and Outbound segments. Nevertheless, due to a lack of information about the end of PSEs in the datasets we used for our experiments, in the rest of the paper we will focus only on the Inbound segments.

2.3 Comparing Events on Different Time-Spans

One of the key problems we had to face for the identification of road segments potentially affected by PSE is how to compare phenomena happening at different times. To better clarify, let us consider an example: a PSE A takes place on a Sunday at 15:00, while a PSE B on a Friday at 18:00. If we notice different traffic trends on the same road segment immediately before A and B, is it due to different hype of the events, or to the rush hour on Friday evening?

In [18], we simply identified, for each day of the week, the time interval with the highest number of events in our dataset, and we excluded all the other information happening outside this interval. This approach has the disadvantage that it forces us to limit the number of events under investigation to those that actually occur within the identified timespan, thus drastically reducing the number of observed events.

To have a more general solution, since we were not able to find in the literature a suitable methodology to compare traffic patterns of the same road in different time instants, we introduce the definition of *Relative Timespan of Interest*, based on the fact that traffic flow follows a regular pattern, that can be modelled, representing the daily Recurring Traffic. Consequently, by definition, any difference between this model and the traffic on a specific day captures the Non-Recurring Traffic. Thus, with the Relative Timespan of Interest, we focus on measuring differences between the traffic model for a given segment on day of the week, and the actual traffic flow on a certain date on that street. If there is a difference, there is a non-recurring influencing factor perturbing the flow on that date.

With this strategy, we can analyse all events for a given weekday, which highly increases the observations in our analysis. However, a clear drawback lies in the traffic modelling that is applied. Unknown events or other influencing factors on traffic (e.g., a thunderstorm) might influence the created models such that they do not represent the traffic behaviour any more, which could lead to a misinterpretation of additional traffic [20].

2.4   The Binary Classification Approach

In order to identify the road segments affected by a PSE, as described in Sect. 2.2, the idea we investigated is to employ a machine-learning technique, searching for correlations between Non-Recurring Traffic and the presence of an event, on each road segment around a venue. This approach is based on the assumption that certain road segments show a different behavior on event days than on non-event days. Therefore, a comparison between the traffic behavior on event days and non-event days is required, limited to the Relative Timespan of Interest of each event. Consequently, a binary classifier can be trained to discriminate road segments of the dataset into positive (Non-Recurring Traffic) and negative (no abnormal traffic behavior) classes, depending on their traffic pattern.

To this aim, we first built a dataset containing, for each road segment within a given radius around a venue, the differences in the traffic behaviour between event days and non-event days. The collection of traffic data is thereby aimed at providing a balanced dataset between traffic on event and non-event days.

A major challenge of this task is to define a time window to compare traffic during event and non-event days. The identification of the time window for event days is straightforward, as we can simply define a time window around the event start time. As for non-event days, we defined the *day-pairs*, i.e. a set of two days, where one is an event day and the other one is a non-event day. The event time on the event day leads to the definition of the time of interest for that *day-pair*, and we choose the same time window from both days. As traffic behaves differently on different days of the week, we need to ensure that both days of the *day-pair* are comparable. To this end, we followed the same *day of week classification* as already used in similar works (e.g. [24]): Monday to Thursday (Mo–Th), Friday (Fr), Saturday (Sa), and Sunday (Su). To create the *day-pairs* and required models for the *Relative Timespan of Interest*, we split our dataset in three sections, for each of these *Day of Week (DoW)* classes:

- Event set (E)
- Non-event set (N)
- Model set (M)

where $E_{DoW} \cup N_{DoW} \cup M_{DoW} = D_{DoW}$ [20]. The event set E contains all event days for that specific venue that we analyse. The non-event set N contains a list of some randomly selected non-event days, whose number is the same of the days in the E set. The model set M contains all non-event days of the database that are not part of N. To generate the dataset, we proceed as follows:

1. We build a traffic model for each *day of week*, by computing the average traffic trend on all the non-event days in M_{DoW}.
2. For each event day in E_{DoW}, we consider the traffic within the *event time window* of 120 min before the start of the PSE. The same time window is also used for one non-event day, randomly chosen from N_{DoW}.
3. We calculate the difference of traffic within such *event time window* between event and non-event days from the model.
4. he resulting *day-pair* is added to the dataset used for the investigation.

To exemplify, let us consider a PSE happening on a Sunday at 17:00. The application of the *event time window* of 120 min, results in considering all the traffic data for Sunday 15:00–17:00. Our approach would add that specific time window to the dataset, together with a randomly selected day among all the Sundays from 15:00–17:00 without events (thus from the N_{Sunday} set). The final *day-pair* would contain information on the difference in traffic from the Sunday model for the event day from 15:00–17:00 and for a randomly selected non-event day on Sunday from 15:00–17:00.

The resulting dataset contains traffic data normalized over defined time periods (we used 5 min slots) for the selected time frames (120 min) in combination with the information of whether an event happened at the specific time. The occurrence of the event (false = no event, true = event) is used as the label for the training/validation of the performances of the classifier. In the end, we obtain a dataset containing, for each road in a specified radius around the venues to analyze, containing the following information:

$$X_1, \ldots X_n, E \tag{1}$$

where $X_1, \ldots, X_n$ are the traffic differences, in percentage, with respect to the model for that day of the week, n is the length in the number of observations for the timespan we are investigating, and E is the event label [20].

Then a classifier can be trained on top of this dataset, potentially in combination with some technique for parameter optimization (e.g. as in [25]), for each road segment.

3   The Experimental Setting

In this section we describe the experiments we conducted to analyse the impact of PSEs on traffic around venues in four different cities in Germany, namely Frankfurt, Dortmund, Berlin, and Cologne.

3.1   Dataset Description

For our analysis we had to deal with three different data sources:

1. A description of the Road Network
2. Traffic Data
3. Information about Planned Special Events.

As for the *Road Network*, we used a commercial automotive solution, based on the Navigation Data Standard (NDS)[3]. Within this data format, each road is split into several segments, whose length may vary from a few meters (mostly at intersections) up to some kilometres. Each road segment has information about its geographical features, the road class (highway, urban, etc.), driving directions and more. The NDS format has been mainly defined to enable future Advanced Driving Assistant Systems (ADAS) and/or Automated Driving solutions. Thus, using it for traffic analysis needs some additional preprocessing steps. Especially the variable length of the segments create problems, since very short segments of the road show huge fluctuation in terms of traffic data. Therefore we filtered our road network to ignore segments shorter than 30 m. We analysed all road segments within a radius of 4000 m around the venues. We did that limitation to reduce calculation times, but the approach is, from a theoretical point, not limited to any maximum radius. The decision for this specific radius came from the experiences we gained during our investigations of PSEs.

The *Traffic Data* we used is a set of speed information for the entire Germany, between 2014 and 2015. It is about traffic information for major road segments, exploiting a table-based referencing system called *traffic message channel (TMC)*, updated twice per minute. Each message contains information for road segments whose current speed fell below 80% of the speed that could be driven under free-flow conditions. As an example, if the usual free-flow speed on a given road segment is 100 km/h, we receive a message whenever the monitored speed goes below 80 km/h. That 80% threshold was defined by the data provider. As long as the monitored speed remains below that threshold, we received continuous updates for that road stretch. Each data package received contains information for all road segments within Germany. We thus defined a

---

[3] http://www.nds-association.org/.

pre-processing pipeline, to parse it and then to map-match it on the NDS road network, using open-source tools. Further details about our pre-processing tool-chain can be found in [26]. Then, for each road segment around a venue hosting PSEs, we checked if there were messages about Non-Recurring Traffic, for event and non-event days. This allowed us to define a labelled dataset for our classifier, containing information about the average of observed delay times in 5 min slots for the selected time spans.

As for the Planned Special Events, we observed four different venues, of which two were focused exclusively on soccer games and the other two hosted different events of all types (see Table 1).

Table 1. List of venues.

Name	City	Type
Commerzbank Arena	Frankfurt (Main)	Soccer
Signal-Iduna-Park	Dortmund	Soccer
Mercedes-Benz Arena Berlin	Berlin	Mixed
LANXESS-Arena	Cologne	Mixed

In total, we observed one season of soccer games in the German First League (*Bundesliga*), which ended up with 12 games per venue. As there were also days included when our traffic dataset was incomplete, we ended up with 11–12 games per venue. For the "mixed" venues, we covered a broad range of events during the time period from 05/2014 to 05/2015 and for each venue, we captured between 50 and 60 events.

3.2   Experimental Procedure

On the above described dataset, we investigated the performances of three classifiers, namely Artificial Neural Networks, K-Nearest Neighbour, and Support Vector Machines (with linear and RBF Kernel), in combination with a simple parameter optimization technique. This because the main goal of this research is to define and validate a general and robust methodology, rather than to find the best possible classification performances. Since the considered classification techniques performed very similarly, in the following we report only the results for the algorithm requiring the least training time, that in our case was the K-Nearest Neighbour. Anyhow, no statistically significant differences were found among the methods. As for the adopted distance metric, we found that the *Dynamic Time Warp (DTW)* was the one able to lead to the best results on our data, most likely due to the robustness of DTW with respect to temporal shifts of similar trends. As a results, in the rest of the paper, we report on the results of the K-NN classifier, using DTW as a distance metric.

To assess the performances on the proposed approach, we applied a leave-one-out cross validation for road segment within the defined radius around each venue, and then we used the standard Information Retrieval metrics, namely Precision, Recall, and F-Measure on the average of the leave-one-out, for each road segment. A low value of Precision indicates high false positives, leading potentially to overestimating the impact range of PSEs. A low Recall means that the system is failing in capturing impacted road segments. To measure the performance of the approach using a single metric, we use *f-measure* that corresponds to the harmonic mean of Precision and Recall.

4   Results and Discussion

In this section we present the results for the four investigated venues, split by venue type.

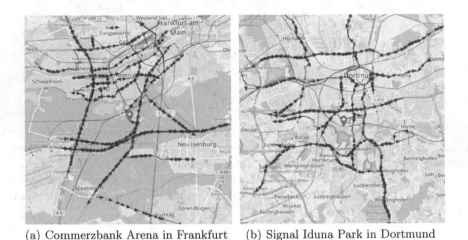

(a) Commerzbank Arena in Frankfurt   (b) Signal Iduna Park in Dortmund

Fig. 1. Road segments with traffic, around the two soccer venues [20].

4.1   Soccer Venues

As for the results about the two analysed soccer venues, considering a radius of 4000 m around the stadiums, we investigated a total of 158 road segments with TMC traffic messages around the Commerzbank Arena in Frankfurt (see Fig. 1(a)), and 131 for the Signal Iduna Park in Dortmund (see Fig. 1(b)). As for the Commerzbank Arena, among these 158 road segments with some traffic incident, 22 showed a detectable correlation with the occurrence of soccer games. Figure 2 shows the results for the *K-NN* model around the Commerzbank Arena for two different *f-measure* thresholds, together with the official parking lots of the stadium.

The road segments with the highest *f-measure* (≥ 0.90 shown in Fig. 2(a)) are those directly leading to the stadium on the highway B44. As all these segments directly lead toward the recommended parking lots and the venue itself, this result seems intuitive. Also the set of road segments around marker **A**, which seem a bit far away from the venue, lead to another parking lot within the area around **A**. From the website of the stadium (own translation): "...within the area of Lyonerstraße, Herriotstraße and Hahnstraße, you can find additional parking...". With a threshold ≥0.85 (Fig. 2(b)), additional road segments become detected (label **B**). They also lead directly to the parking lot close to the stadium. Below 0.85, some misclassifications appears, with unexplainable involved segments.

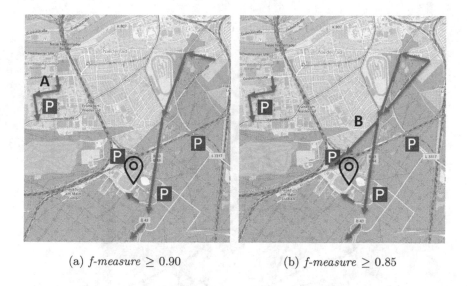

(a) *f-measure* ≥ 0.90   (b) *f-measure* ≥ 0.85

Fig. 2. Results for venue: Commerzbank Arena Frankfurt.

As for the Signal Iduna Park, among the 131 road segments with some traffic delay, 18 showed a remarkably different behaviour during soccer games. Figure 3 shows the results for the K-NN model, given two *f-measure* thresholds. Here, the segments with the highest *f-measure* are those pointing towards the stadium. Indeed, to reach the venue, drivers have to cross the intersection at label **A** and then proceed toward the stadium. As this intersection serves multiple incoming segments, these results seem reasonable. The result for the segments around the label **B** are also pretty clear, since, as described on the official website[4] the highway B54 (the segment around B is a part of it) is suggested as an incoming route for people arriving from the south. As a confirmation, the detected segments on the B54 end at the exit towards the stadium. Below the threshold of 0.80, there are some questionable results, probably misclassified. It is also worth noting that the Signal Iduna Park in Dortmund is the biggest soccer stadium in Germany, with a capacity of more than 80,000 visitors. This could be a reason why event traffic is very explicitly noticeable in this area.

---

[4] https://www.signal-iduna-park.de/.

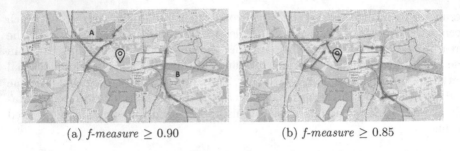

(a) *f-measure* ≥ 0.90   (b) *f-measure* ≥ 0.85

Fig. 3. Results for venue: Signal Iduna Park Dortmund.

Fig. 4. Selected roads within a radius of 4000 m around the Mercedes-Benz Arena in Berlin (black marker). In green all the other venues in the area. (Color figure online)

4.2 Mixed Venues

The first "mixed" venue we investigated is the Mercedes-Benz Arena in Berlin. During the time of our investigation, and considering also some problems we had in collecting data around this venue, we ended up with 48 events for which both event and traffic information were available. This is a significantly higher amount, if compared with the 11–12 games observed at the soccer venues. In total, we collected a total of 285 roads with some traffic even reported by TMC, within the 4000 m radius, as shown in Fig. 4.

The highest *f-measure* observed for this venue is as little as 0.72 (on just segment), being by far lower than those of the soccer venues discussed before.

This most correlated road (label **A** in Fig. 5(a)) is directly in front of the venue, similarly to observations at the other venues. Reducing the threshold ≥0.65, we have five segments (see Fig. 5(b)). Some of them are plausible, others not that much, like those close to the label **B** in Fig. 5(b). Below a threshold of 0.65, the results become inconsistent and obvious misclassifications appear.

(a) *f-measure* ≥ 0.70   (b) *f-measure* ≥ 0.65

Fig. 5. Results for venue: Mercedes Benz Arena Berlin.

A possible explanation for these low results lies in density of venues within the considered radius, which may interfere with the observations for this particular one. Indeed, as shown in Fig. 4, we found 77 venues within a radius of 4000 m around the Mercedes-Benz Arena, which could be anyhow an underestimate. Events happening at the same time in different venues could impact on the road network during time spans that we counted as non-event times. For example, on Thursday 28/05/2015, which is counted as a non-event day for the analysis, we observed severe traffic congestion in the area. On that day, three different events happened within 1000 m from the Mercedes-Benz Arena. Two of them were concerts (Rhodes played at *Berghain* and Leslie Clio at the *Postbahnhof Club*) and one cultural event at the *Kriminaltheater*. Each one of these could have caused the observed delay. Due to these interferences, a clear separation between event and non-event days, exclusively relying on traffic observations, is difficult to achieve. Another possible explanation lies in the event type diversity. In contrast to the first venues that mostly hosted soccer games, in this venue also mixed types of event categories (concert, culture, musical&show, and sport). It is recognized that those categories of events show a different impact on traffic

[17]. The Fig. 6 shows the sum of the delays over all the roads in a radius of 1000 m from the venue, for the different event categories, which are sensibly different [20].

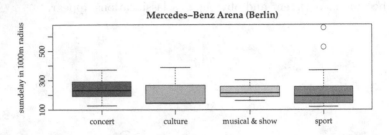

Fig. 6. Avgdelay variation over time attributes within a radius of 1000 m around the Mercedes-Benz Arena Berlin [20].

Similarly to the Mercedes-Benz Arena in Berlin, also the LANXESS Arena in Cologne shows poor classification scores. In this venue, we got traffic messages for 160 road segments within a radius of 4000 m, over 33 events. The highest observed *f-measure*, for just one segment, is 0.71. In Fig. 7 we report the results the threshold ≥0.65. The road segments around the label **A**, very close to the Arena, are those with an *f-measure* ≥0.7. All the others are in the range 0.65–0.69. In total, we find five roads ≥0.65. In Cologne, the venue density is still very high, even if not at the same level of Berlin, as there are fewer venues in the immediate surroundings of the Arena. However, north of it, there is the

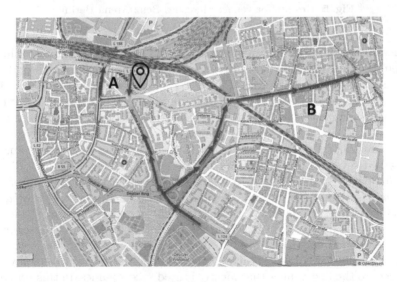

Fig. 7. Venue: LANXESS-Arena Cologne (green marker). *F1-measure* ≥0.65. (Color figure online)

Messegelände Köln, also hosting many events. In addition, most of the venues are in the inner city (west of the arena). Since there are only a few bridges to cross the river, to go to the inner city, events happening there could also influence traffic in a wider area. This arena hosted events of four different categories (concert, misc, musical&show, and sport). Whereas all of them showed severe variations in the observed traffic behaviour, the *concert* category showed the highest fluctuations. These observed variations between events drastically reduce the performance of the classifier and explain the poor results for this venue [20].

5   Conclusions and Future Work

Improving mobility and reducing the social end economic negative effects of traffic congestions is one of the greatest challenges for the Intelligent Transportation Systems community. Within the data-driven ITS topic, lots of research has been devoted to learn the impact of different factors on mobility. However, to the best of our knowledge, only few research attention has been dedicated to the impact on traffic of Planned Special Events. To fill this gap, we have presented an approach, based on any robust binary classifier, to automatically identify road segments in inner city scenarios that are impacted by PSEs. That method is an enabling technology to (I) capture the impact of PSEs on traffic, and (II) learn its behaviour and predict it, for future smarter route calculation engines. We empirically evaluated our approach on four different venues in Germany (Frankfurt, Dortmund, Berlin, and Cologne). Results show that our approach is able to correctly identify road segments that people tend to use on their way to the venue for soccer stadiums. In this scenario, some road segments show a F-measure up to 0.97 and a further visual evaluation of the are confirms the goodness of the results. On the other hand, for the two venues hosting mixed events, results are less positive, probably due to the complexity of the road network and the influence of other venues in the considered areas. This work is a first step towards the goal to predict the impact of PSEs on traffic. Indeed, a lot of research is still needed in this field. For example, to understand the reasons of the variability among the event categories, with the goal to model these differences in behaviour in advance, more information sources about those events could be exploited. As a clear extension of this research, we are planning to feed the classifier with information about PSEs coming from other data sources (e.g. social media, news events, etc.). Another aspect that can be further investigated is the use of more advanced machine learning techniques, like deep learning.

References

1. Manyika, J., et al.: Big data: the next frontier for innovation, competition, and productivity. Technical report, McKinsey Global Institute (2011)
2. Kwon, J., Mauch, M., Varaiya, P.: Components of congestion: delay from incidents, special events, lane closures, weather, potential ramp metering gain, and excess demand. Transp. Res. Rec.: J. Transp. Res. Board **1959**(1), 84–91 (2006)

3. Horvitz, E., Apacible, J., Sarin, R., Liao, L.: Prediction, expectation, and surprise: methods, designs, and study of a deployed traffic forecasting service. In: Twenty-First Conference on Uncertainty in Artificial Intelligence (2005)

4. Pan, B., Demiryurek, U., Shahabi, C., Gupta, C.: Forecasting spatiotemporal impact of traffic incidents on road networks. In: 2013 IEEE 13th International Conference on Data Mining (ICDM), pp. 587–596, December 2013

5. Zhang, X., Rice, J.A.: Short-term travel time prediction. Transp. Res. Part C: Emerg. Technol. **11**(34), 187–210 (2003)

6. Ishak, S., Al-Deek, H.: Performance evaluation of short-term time-series traffic prediction model. J. Transp. Eng. **128**(6), 490–498 (2002)

7. Wu, C.-H., Ho, J.-M., Lee, D.: Travel-time prediction with support vector regression. IEEE Trans. Intell. Transp. Syst. **5**(4), 276–281 (2004)

8. Van Lint, J.W.C.: Online learning solutions for freeway travel time prediction. IEEE Trans. Intell. Transp. Syst. **9**(1), 38–47 (2008)

9. Park, D., Rilett, L., Han, G.: Spectral basis neural networks for real-time travel time forecasting. J. Transp. Eng. **125**(6), 515–523 (1999)

10. Vanajakshi, L., Rilett, L.: A comparison of the performance of artificial neural networks and support vector machines for the prediction of traffic speed. In: 2004 IEEE Intelligent Vehicles Symposium, pp. 194–199, June 2004

11. Vlahogianni, E.I., Karlaftis, M.G., Golias, J.C.: Short-term traffic forecasting: where we are and where we are going. Transp. Res. Part C: Emerg. Technol. **43**, 3–19 (2014)

12. TomTom International: White paper - how TomTom's hd traffic and iq routes data provides the very best routing. Technical report, TomTom International (2009)

13. Dunn, W., Latoski, S.: Managing travel for planned special events. In: Institute of Transportation Engineers (ITE) 2003 Technical Conference and Exhibit, no. CD-020 (2003). http://trid.trb.org/view.aspx?id=646244

14. Skolnik, J., Chami, R., Walker, M.: Planned special events-economic role and congestion effects. Technical report, Federal Highway Administration, US-DOT (2008)

15. Lei-Lei, D., Jin-Gang, G., Zheng-Liang, S., Hong-Tong, Q.: Study on traffic organization and management strategies for large special events. In: 2012 International Conference on System Science and Engineering (ICSSE), pp. 432–436, June 2012

16. Kwoczek, S., Di Martino, S., Nejdl, W.: Predicting traffic congestion in presence of planned special events. In: Proceedings of the Twentieth International Conference on Distributed Multimedia Systems, DMS 2014, pp. 357–364 (2014)

17. Kwoczek, S., Di Martino, S., Nejdl, W.: Predicting and visualizing traffic congestion in the presence of planned special events. J. Vis. Lang. Comput. **25**(6), 973–980 (2014)

18. Kwoczek, S., Di Martino, S., Nejdl, W.: Stuck around the stadium? An approach to identify road segments affected by planned special events. In: 2015 IEEE 18th International Conference on Intelligent Transportation Systems (ITSC), pp. 1255–1260. IEEE (2015)

19. Pereira, F.C., Rodrigues, F., Polisciuc, E., Ben-Akiva, M.: Why so many people? Explaining nonhabitual transport overcrowding with internet data. IEEE Trans. Intell. Transp. Syst. **16**(3), 1370–1379 (2015)

20. Kwoczek, S.: Enhanced mobility awareness: a data-driven approach to analyze traffic under planned special event scenarios. Ph.D. Dissertation, Institutionelles Repositorium der Leibniz Universität Hannover, Hannover (2018)

21. Matherly, D., Murray-Tuite, P., Wolshon, B., Pande, A., Wolshon, B.: Traffic management for planned, unplanned, and emergency events. In: Traffic Engineering Handbook, pp. 599–636. Wiley (2015)

22. Dowling, R., Skabardonis, A., Carroll, M., Wang, Z.: Methodology for measuring recurrent and nonrecurrent traffic congestion. Transp. Res. Rec.: J. Transp. Res. Board **1867**, 60–68 (2004)
23. Anbaroglu, B., Heydecker, B., Cheng, T.: Spatio-temporal clustering for nonrecurrent traffic congestion detection on urban road networks. Transp. Res. Part C: Emerg. Technol. **48**, 47–65 (2014)
24. Richter, F., Di Martino, S., Mattfeld, D.C.: Temporal and spatial clustering for a parking prediction service. In: 2014 IEEE 26th International Conference on Tools with Artificial Intelligence (ICTAI), pp. 278–282. IEEE (2014)
25. Corazza, A., Di Martino, S., Ferrucci, F., Gravino, C., Sarro, F., Mendes, E.: Using tabu search to configure support vector regression for effort estimation. Empirical Softw. Eng. **18**(3), 506–546 (2013)
26. Kwoczek, S., Di Martino, S., Rustemeyer, T., Nejdl, W.: An architecture to process massive vehicular traffic data. In: 2015 10th International Conference on P2P, Parallel, Grid, Cloud and Internet Computing (3PGCIC), pp. 515–520. IEEE (2015)

Designing a Query Language
Using Keyword Pairs for Spatial
and Temporal Search

Yuanyuan Wang[1]($\boxtimes$)(iD), Panote Siriaraya[2], Haruka Sakata[2], Yukiko Kawai[2,3](iD),
and Keishi Tajima[4]

[1] Yamaguchi University, 2-16-1 Tokiwadai, Ube, Yamaguchi 755-8611, Japan
y.wang@yamaguchi-u.ac.jp
[2] Kyoto Sangyo University, Motoyama, Kamigamo, Kita-ku, Kyoto 603-8555, Japan
spanote@gmail.com, kawai@cc.kyoto-su.ac.jp
[3] Osaka University, 5-1 Mihogaoka, Ibaraki, Osaka 567-0047, Japan
[4] Kyoto University, Yoshida-Honmachi, Sakyo-ku, Kyoto 606-8501, Japan
tajima@i.kyoto-u.ac.jp

Abstract. Our goal is to design a query language based on two keywords for spatial and temporal search by using a single textual query in an intuitive way. This language as a form of syntactic sugar can express complex spatio-temporal queries which include conditions on range, direction, time length and size which are difficult to express through textual queries in conventional keyword-based search systems. To express these conditions, our proposed language introduces 12 spatio-temporal operators such as arithmetic and directional operators which enables users to combine and manipulate spatial and temporal areas. Also, we use "space characters" (the space-key) between keywords which are used to express the geographical distance or time-length between matching objects in an intuitive way for general users. In this paper, we provide an overview of our proposed search system where we can retrieve maps and web documents to highlight how the query language could be put into practice by using complex spatio-temporal queries. Finally, we discuss the results of a user study carried out to evaluate the potential usefulness of our proposed search system.

Keywords: Query language · Spatial and temporal search ·
Space-key search · Spatio-temporal operators

1 Introduction

Although nonverbal search services continue to increase in modern society, have you ever been to conscious about using a space-key ($\sqcup$) while typing search queries? Probably "no" because space characters do not usually have important meaning in keyword search other than in the conjunction of the components. We propose to give a valuable and useful role to space characters in search queries

© Springer Nature Switzerland AG 2019
Y. Kawai et al. (Eds.): W2GIS 2019, LNCS 11474, pp. 118–135, 2019.
https://doi.org/10.1007/978-3-030-17246-6_10

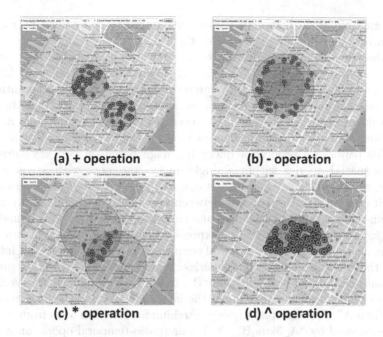

(a) + operation   (b) - operation

(c) * operation   (d) ^ operation

Fig. 1. Example of map search using proposed operators.

based on two keywords. Space characters are useful and powerful for expressing
spatio-temporal relationship in queries for maps, images, videos, and web pages,
while it is easy to use and intuitive for users.

While, simple keyword-based queries have been widely adopted in search
systems because of their ease of use, they are not good at expressing spatial
and temporal relationships. Such queries are difficult to find appropriate maps,
images, and videos, when more complex requests are needed. For example, when
we travelling by car, to find a famous pizza parlor within 100–500 m from our
current position on a map service supporting keyword queries, we need four
transactions: (1) search for pizza parlors on the map by using a keyword query,
(2) also identify current position on the map, (3) limit the query result to those
which are within 500 m from the current position, and (4) also exclude those
which are within 100 m from the current position. Queries including more complex conditions are even harder to execute in a single transaction. Such queries
include: ($Q1$) "Find all pizza parlors which are within 500 m from our current
position and within 500 m from location B", ($Q2$) "Find all pizza parlors which
are within the range of 100–500 m from my current position", ($Q3$) "Find all
pizza parlors that are within 500 m from my current position which are also
within 500 m from location B", and ($Q4$) "Find all pizza parlors located to
the north of and are also within 500 m from my current position". As shown
in these examples, it is difficult to process complex search requests using simple keyword-based queries. Systems only supporting such keyword-based queries

require multiple steps including non-textual interactions or need to use very complicated query languages to process them.

On the other hand, there have been much research on spatial logic or algebra. In a spatial database of PostGIS[1], the spatial search task with location queries can be run in SQL, and they can represent complex spatial conditions in queries, they require users to learn and understand the programming-language-like syntax, and as a result, they are too complicated for general users in many applications. They are, therefore, impractical for the use in such systems.

In this paper, we propose a query language based on two keywords that expresses spatial queries within a single query statement with a concise and intuitive syntax so that general users can easily specify complex queries. Our approach uses the length of spaces between two keywords in queries to express conditions on geographical distances and time-length between objects matching the two keywords. This allows us to express queries as a form of syntactic sugar including complicated spatio-temporal conditions in a very simple and intuitive way. In the conventional systems, queries "A␣B" and "A␣␣␣␣␣␣B" return the same results containing both "A" and "B". In our query language however, these two queries have a different meaning: the latter means "Search the 6 nearest B objects from A". For example, a query searching for B within 5 km from A on the map is expressed by "A␣5km␣B". A similar spatio-temporal operation can also be used for retrieving documents and video. For example, a query searching for a document containing "B" within 5 sentences from "A" is expressed by "A␣5 sentences␣B".

Our language also introduces 12 operators including set operators and spatio-temporal operators to express queries including conditions on directions, ranges, angles and time. Figure 1 shows an example of map search using operators ([+] [-] [*] [^]). For example, $Q1$ shown above can be expressed by using a union operator ([+]) as follows: (A␣800m) + (B␣800m)␣pizza parlor, which aims to identify all the pizza parlors found within the combined region which is within 800 m from A and within 800 m from B (see Fig. 1(a)). $Q2$ can be expressed by using a range operator ([-]) as follows: (A␣800m) − (B␣600m)␣pizza parlor, which aims to identify all the pizza parlors located within 800 m from A and excluding those which are located within 600 m from B (see Fig. 1(b)). $Q3$ can be expressed by using a set operator ([*]) as follows: (A␣0.5km␣pizza parlor) * (B␣0.5km␣pizza parlor), which aims to identify all pizza parlors located within 0.5 km both from A and from B (see Fig. 1(c)) and $Q4$ can also be expressed by using directional condition operators ([^] or [θ]) as follows: A␣^500m␣pizza parlor, which aims to identify all pizza parlors located to the north of and are within 500 m from A (see Fig. 1(d)). We also introduce the property operator ([$]), which is used to test properties of objects matching to query keywords. For example, a query "current position␣500m$20people␣pizza parlor" retrieves pizza parlors within 500 m which can cater to 20 people.

The remainder of this paper is structured as follows. In Sect. 2, we discuss previous research which has been carried out related to map, text and video search.

---

[1] https://postgis.net/.

Afterward, in Sect. 3, we provide a definition of spatio-temporal operators and provide examples of how they could be applied to query data. Section 4 describes the structure and components of a spatio-temporal search system. Section 5 provides application examples of an implemented spatio-temporal search system. Section 6 discusses our proposed search method with a user study. Finally, in Sect. 7, we conclude this paper and discuss future works.

2 Related Work

2.1 Map Search

Most of the current major location services such as Google Maps or Bing Maps focus on finding certain locations within a specified geographical area or the best routes (e.g., shortest distance, most economical) matching given query keywords. For example, when you want to find restaurants in Kyoto station, a query "restaurant␣in␣Kyoto station" could be used to find restaurants in Kyoto station, and when you want to find supermarkets around Tokyo station, the query "supermarket␣near␣Tokyo station" could be used to find supermarkets around Tokyo station. In regards to finding appropriate routes, there have been many studies aimed at helping users locate simple and memorable [3,19], comfortable [10,15,17], safe [5,9,12], or aesthetically pleasing routes [16] in map services, as well as studies which seek to personalize the search results, by identifying locations which better match the latent interests of users [11,21].

These search systems are often designed for general users and they only utilize keyword matching algorithms to process user search queries. For example, users can simply input a query such as "supermarket in Tokyo station" to search for supermarkets around Tokyo station. However, these systems cannot satisfy users when they wish to retrieve more complex and precise search requests. Although several API systems, i.e., Bing Maps APIs[2] and Google Maps Platform[3] provide access to more advanced features of map search systems, they are often limited to single-process tasks (e.g., finding places within a specific distance, geocoding a certain location name). Therefore, in this paper, we propose a novel spatio-temporal query language as a form of syntactic sugar that can express conditions on distance and time-length between objects matching the query keywords in an intuitive way for general users, and we also show the demonstration for map search. For example, we can issue a query "Tokyo station␣500m␣supermarket" to find supermarkets within 500 m from Tokyo station.

2.2 Spatio-Temporal Search for Text and Video

Proximity operators are used in some document search systems or Web search engines to narrow search results by limiting them to those that have certain query keywords placed within a specified distance or in the specified order. For example,

---

[2] https://www.microsoft.com/en-us/maps/choose-your-bing-maps-api.
[3] https://cloud.google.com/maps-platform/.

ScienceDirect and Scopus utilize a proximity operator (W/n) which retrieve documents where query keywords appear within n words [4]. For example, a query "microscopy W/3 gfp" could be used to find literature containing the word "microscopy" that is within 3-word lengths of the word "gfp". Oracle Text utilize a proximity operator (near) which requires the query keywords to appear within a default number of word distances [8]. For instance, a query "near((dog, cat), 6)" could be used to search for all documents which include the word "dog" that is within 6-word lengths of the word "cat". In our case, we express constraints on the distance between query keywords (or objects matching query keywords) by using space characters (␣) between the query keywords, allowing general users to easily express queries that include spatio-temporal conditions.

Video interval operators such as union, intersection, concatenation, and their set-variants have been devised to allow users to programmatically edit videos [7,20]. Although researchers have defined several interval operations to compute new intervals from existing intervals, the new intervals cannot satisfy users' search requests as they lack certain types of algebraic operators such as difference. However, given a set of fragmentarily indexed video shots, these operations generally produce fragmentary intervals and thus cannot always produce appropriate intervals which users generally intend to find. Pradhan et al. also proposed interval operators called *glue join* operations for composing longer intervals from a set of short annotated intervals [13,14,18]. These studies focus on using operators to compose semantically meaningful video intervals and determines how to efficiently evaluate them. On the other hand, the purpose of our research is to design a query syntax that can express spatio-temporal conditions in an easy and intuitive way.

3   Spatio-Temporal Query Language

In this research, we define 12 operations including set operations and spatio-temporal operations by text for use in search operations. We mainly show examples taken from map search tasks, but we also show examples from Web search and video search tasks.

Definition 1: The syntax of the most primitive unit of the spatio-temporal query is defined as follows: A␣spatio-temporal length␣α. A and α are keywords, with A denoting the location of the origin for a spatio-temporal search for the object with the property α. This permits us to use more than two contiguous spaces from the first appearing space (␣) by using the space-key. The part corresponding to "spatio-temporal length (spatio-temporal operator)" should follow after the last space (␣) in those contiguous spaces. When continuous spaces are used, the spatio-temporal length would represent the N nearest locations with the property α, where N is represented by the number of continuous spaces.

    (Example): A␣800m␣α

        denotes a query statement to identify the α objects which exist inside the region 800 m from the origin point A.

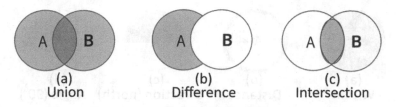

Fig. 2. Set operators.

Table 1. Spatial operators by using the space-key

Operator	␣*	␣^	␣_	␣@	␣[x-y]	␣$	␣#
Processing	Surrounding	Direction (north/up)	Direction (south/down)	Angle	Range	Size	Time

(**Example**): $A_{\sqcup\sqcup\sqcup}\alpha$
   denotes a query statement to identify the 3 nearest α objects from the origin point A.

Definition 2: The spatio-temporal length used in the syntax above are represented by a unit distance (e.g., 100 m, 1 sentence, etc.). For example, the unit distance "800 m" could be used following the $\sqcup$ spatio-temporal operator.

Definition 3: The keywords (e.g., A and α) used in the primitive unit of the spatio-temporal query would be encapsulated within a double quotation mark (e.g., "Tokyo tower" or "Grand Central Terminal, New York" for A or "pizza shop" or for α). In addition, various spatial, directional and distance operators could be used to impose conditions when conducting a spatio-temporal search.

Definition 4: Each primitive spatio-temporal search unit could be combined with other units through the use of spatial, directional, and distance operators in a mathematical expression format as follows: $SQ^1 \cup SQ^2 \cup ... = SQ^+$.

   (**Example**): $(A_\sqcup 800m_\sqcup\alpha) + (B_\sqcup 500m_\sqcup\alpha)$
   denotes a query to identify the union of the α objects within 800 m from point A and the region within 500 m from point B.

3.1 Set Operators

The standard set operators can also be used as spatial-temporal operations for the query unit defined previously [1,6]. These include the union [+], difference [−], and intersection [*] operators. Users could use such operations to manipulate the spatial region they wish to search into. Examples of queries including these operators are shown below:

Union calculation (Ex.1): $(A_\sqcup 3km_\sqcup\alpha) + (B_\sqcup 3km_\sqcup\alpha)$
   denotes the union of the spatial region *within* 3 km from point A AND the spatial region *within* 3 km from point B (see Fig. 2(a)).

Difference calculation (Ex.2): $(A_\sqcup 3km_\sqcup\alpha) - (B_\sqcup 3km_\sqcup\alpha)$
   denotes the spatial region *within* 3 km from point A which is NOT *within* 3 km from point B (see Fig. 2(b)).

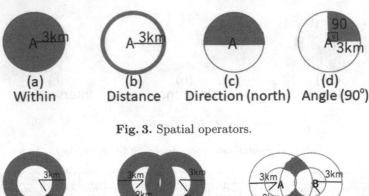

Fig. 3. Spatial operators.

Fig. 4. Range operators.

Intersection calculation (Ex.3): $(A_\sqcup 3km_\sqcup\alpha) * (B_\sqcup 3km_\sqcup\alpha)$
denotes the spatial region that is *within* 3 km from point A which is ALSO *within* 3 km from point B (see Fig. 2(c)).

All set operators (union [+], difference [−], intersection [*]) can be used to search for objects with the properties identified in the query unit. For example, the aforementioned $(A_\sqcup 3km_\sqcup\alpha) * (B_\sqcup 3km_\sqcup\alpha)$ query would search for objects with the property α which is located within the spatial region that is the result of the intersection between 3 km from points A and B. Note that only union [+] and intersection [*] can be used between two search objects if the search objects have different types (e.g., restaurant and hotel).

3.2   Spatial Operators

Table 1 shows the 7 spatial operators which could be used to further denote distance and direction within our proposed spatial-temporal search queries. Examples of four expressions which use these operators (within, distance, direction, and angle) are described below:

Within operation (Ex.4): $A_\sqcup 3km_\sqcup\alpha$
retrieves the α objects which are *within* 3 km from point A (Fig. 3(a)).
Distance operation (Ex.5): $A_\sqcup *3km_\sqcup\alpha$
retrieves the α objects that are 3 km *away from* point A (Fig. 3(b)).
Direction operation (Ex.6): $A_\sqcup \hat{} 3km_\sqcup\alpha$
retrieves the α objects which are to the *north* of, and *within* 3 km from, point A (see Fig. 3(c)).
Angle operation (Ex.7): $A_\sqcup 3km@90_\sqcup\alpha$
retrieves the α objects which exist within 3 km in the 90° counterclockwise direction from point A (i.e., is east of point A) (see Fig. 3(d)).

Table 2. Time operators by using the space-key

Operator	⊔#*	⊔#<	⊔#>	⊔#[x-y]	⊔#$
Processing	Around	Before	After	Range	Current

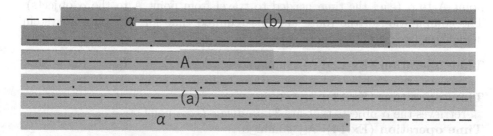

Fig. 5. Example of document extraction by Web search.

3.3 Range Operators

Next, we show three examples of the range operations shown in Table 1.

Range operation (Ex.8): $A_⊔[3km-1km]_⊔\alpha$
  retrieves the α objects in the spatial region from 1 km to 3 km from point A (see Fig. 4(a)). The same condition can also be expressed by the following query, which uses the "difference" operator $[-]$: $(A_⊔3km_⊔\alpha) - (A_⊔1km_⊔\alpha)$.

Range operation (Ex.9): $(A_⊔[3km-1km]_⊔\alpha) + (B_⊔[3km-1km]_⊔\alpha)$
  retrieves the α objects which exist inside the spatial region from 1 km to 3 km either from point A and from point B (see Fig. 4(b)).

Range operation (Ex.10): $(A_⊔[3km-1km]_⊔\alpha) * (B_⊔[3km-1km]_⊔\alpha)$
  retrieves the α objects which exist inside the spatial region from 1 km to 3 km both from point A and from point B (see Fig. 4(c)).

3.4 Size Operators

Below is an example query including the size operator shown in Table 1. The size operator $[\$]$ extracts the size of the corresponding property of object α and uses it as a unit of measure (e.g., 1 city block $= 0.5$ km). This operator is placed immediately after the corresponding object size.

Size operation (Ex.11): $A_⊔3km\$10units_⊔\alpha$
  retrieves the α objects that have a size of 10 units *within* 3 km from point A.

3.5   Time Operators

Time operator [#] shown in Table 1 can be used to formulate time-based search queries. Table 2 shows five different time operators which could be used. For the map search, the time operators express conditions on time distance from point A to α (e.g., the time needed to travel from point A to the α objects). The computation of the route and the required time for the route is system dependent (see Sect. 4 for more details).

Time operation (Ex. 12): $A_\sqcup \#3min_\sqcup\alpha$
retrieves the α objects that are *within* 3 min from point A.
Time operation (Ex. 13): $A_\sqcup \#*3min_\sqcup\alpha$
retrieves the α objects that are 3 min *away from* point A.
Time operation (Ex. 14): $A_\sqcup \hat{\ }\#3min_\sqcup\alpha$
retrieves the α objects that are to the north of, and *within* 3 min of point A.
Time operation (Ex. 15): $A_\sqcup \#3min@90_\sqcup\alpha$
retrieves the α objects that are *within* 3 min from point A and ALSO in the rotated 90° counterclockwise from the east direction of point A.

3.6   Applied to Web and Video Search

In addition to the map search, our proposed spatio-temporal query language could also be applied to Web search and video search.

Application Examples for Web Search. The spatio-temporal query language expresses conditions to extract documents from point A to the α objects.

Within operation: $A_\sqcup 3lines_\sqcup\alpha$
retrieves the lines including the α objects which are *within* the third line from point A (see Fig. 5(a)).
Range operation: $A_\sqcup *3lines_\sqcup\alpha$
retrieves the lines including the α objects that are 3 lines *away from* point A (see Fig. 5(a)(b)).
Direction operation: $A_\sqcup \#<3lines_\sqcup\alpha$
retrieves the lines including the α objects which are *before* 3 lines from point A (see Fig. 5(b)).

Here, the α object can also be used as an extension (e.g., .png) or a link.

Direction operation: $A_\sqcup \#<3lines_\sqcup.jpg$
retrieves the .jpg files (images) which are *before* 3 lines from point A.
Direction operation: $A_\sqcup \#<3lines_\sqcup URL$
retrieves the URL links which are *before* 3 lines from point A.

In addition to searching the line, a paragraph could also be searched. Furthermore, it is possible to perform a search that takes visual direction into account by using DOM analysis. This allows a visual search of where on the page the object exists.

Fig. 6. Example of interval extraction by video search.

Direction operation: A␣^3lines␣.jpg
retrieves the .jpg files (images) which are *above* of and is *within* 3 lines from point A.

Range operation: A␣[@180-@90]␣.jpg
retrieves the .jpg files (images) which layout in the position from the rotated 90° counterclockwise to the rotated 180° counterclockwise from the east direction (left side) of point A.

Application Examples for Video Search. The spatio-temporal query language expresses conditions on video frame extraction from point A to the α objects.

Time operation: A␣#3min␣α
retrieves the α video frames that are *within* 3 min from point A.

Time operation: A␣#*3min␣α
retrieves the α video frames that are 3 min *away from* point A.

Time operation: A␣#<3min␣α
retrieves the α video frames which are *before* 3 min from point A.

Furthermore, it is possible to perform a search that takes visual direction into account by video frame analysis. This allows a visual search of where on the video it exists.

Direction operation: A␣^3min␣α
retrieves the α video frames which are *above* of and *within* 3 min from point A.

Direction operation: A␣[@270-@90]␣α
retrieves the video frames contain α which exist in the screen from the rotated 90° counterclockwise to the rotated 270° counterclockwise from the east direction of point A (see Fig. 6).

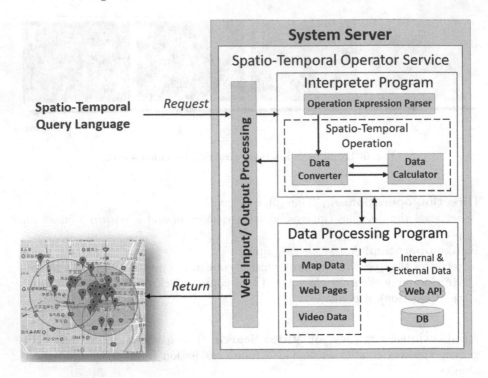

Fig. 7. Overview of the spatio-temporal search system.

4   Spatio-Temporal Search System

In this section, we explain the structure of our proposed spatio-temporal search system (see Fig. 7). The system consists of three main components: (1) a Web Input/Output component that processes user requests and output them to the appropriate format; (2) an interpreter component that parses and processes queries including the spatio-temporal operators; and (3) the data processing program component which is developed by the system developer to link the spatio-temporal search system to appropriate data sources.

4.1   Web Input/Output Processing

By using our system, a client sends an HTTP request to the server with details of the query including spatio-temporal operators and requested data type (map, web page, video) as API parameters. For example, the client retrieving map information by a spatial query $A_{\sqcup}\hat{}3km_{\sqcup}\alpha$ sends the following GET request to the server: www.hoge.com/map/?q=A++^3km+α (or l=A&space=3km&q=α).

The data type and the query specified by the client is then passed to the interpreter and the data processing components. These components would parse the query, process the request, and send the results back to the Web In/Output Processing component which would transmit the results back to the client as an HTTP response in a data format such as JSON or XML format for API.

4.2 Interpreter and Data Processing

The role of the interpreter component is to process the spatial-temporal operators sent as the request from the users. This component consists of a query parser, a spatio-temporal data converter, and a spatio-temporal data calculator. For the parser, the role is to analyze the user query and determine the appropriate operations and procedures to process it. For example, when users input the following query: $(A_\sqcup{}^3km_\sqcup\alpha * B_\sqcup{}^3km_\sqcup\alpha) + (C_\sqcup{}^5km_\sqcup\beta + D_\sqcup{}^1km_\sqcup\beta)$. .

It would be processed by the parser into the following steps:

Var1 $= A_\sqcup{}^3km_\sqcup\alpha * B_\sqcup{}^3km_\sqcup\alpha$ (step 1)
Var2 $= C_\sqcup{}^5km_\sqcup\beta + D_\sqcup{}^1km_\sqcup\beta$ (step 2)
Result $=$ Var1 + Var2 (step 3)

Next, these steps would then be processed by the interpreter. Each spatial variable is sent to the data converter to convert the elements (i.e., $A_\sqcup{}^3km$ or $B_\sqcup{}^3km$) in the query to spatial regions which represents the correct distribution of those elements. The conversion program would access information provided by the data processing component to calculate the appropriate regions. For example, when processing the element "$A_\sqcup{}^3km_\sqcup shops$" for a map, the data processing component would calculate the geographical location of point A as well as the geographical locations of shops within a 3 km radius. If the system developer has specified Google API as the data source, the corresponding data would be obtained through a Web request sent to the API servers. Alternatively, if the developer specified an internal SQL database as the data source, the corresponding data are obtained through an SQL request. The same process would also be carried out for the other elements "$B_\sqcup{}^3km_\sqcup\alpha$", "$C_\sqcup{}^5km_\sqcup\beta$", and "$D_\sqcup{}^1km_\sqcup\beta$".

After the data has been converted, the spatial operators are then processed. For example, if the request query contains the intersection operator [*] in step 1, it would calculate the spatial region which is the overlap between the converted $A_\sqcup{}^3km$ and $B_\sqcup{}^3km$ regions. After all the calculations have been completed, the result is sent back to the client in the appropriate data type (JSON or XML, etc.) as specified by the data processing component.

5 Demonstrative Applications

A prototype of the spatio-temporal search system[4] was implemented as a RESTFUL Web service using Node.js. The current system supports spatial map search, with the input being the requested as a spatial query (an HTTPS GET request) and the output being an array of locations which match the spatial query (returned using the JSON data format). The operation expression within the spatial query was parsed using the Shunting-yard algorithm. Google Maps API was used in the data processing proportion to identify the various locations specified in the primitive spatial query unit (e.g., "Times Square") and their

---

[4] http://yklab.kyoto-su.ac.jp/~sakata/spatialQueryDemo/.

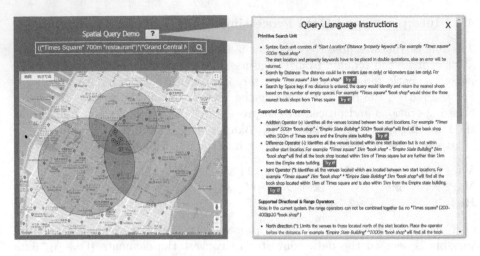

Fig. 8. Spatial query language demo application[4].

geographical positions. The system could also later be easily adapted to utilize other data sources such as Open Street Map data or a customized SQL database as well.

5.1 Applications of Map Search

To highlight how the system could be useful in practice, a number of Web applications were created which utilized our proposed spatial search query language. The first application was a Web interface for our search system which users could use to test the query language or search for locations using the spatial query language. Users would be able to use the spatial, range, and directional operations described in Sect. 3 as well as mathematical expressions such as brackets to compose their search queries. After clicking the search button, the system would send the user's query to the search system server and would then return the search results received from the server onto the map. For example, Fig. 8 shows the results of the query: (("Times Square" 700m "restaurant") * ("Grand Central" 1km "restaurant")) * ("Pennsylvania   station" 1km "restaurant"), which aims to identify all restaurants located within 700 m of Times Square and 1 km from Grand Central and Pennsylvania station. One potential use-case for such a query is to create a "meet up feature", which would identify potential meeting places for three users based on their starting locations. For example, when one user works near Times Square and the other near Grand Central and the final near Pennsylvania station and the system would need to find a restaurant that is equally near to all three of their workplaces for them to meet for lunch. The Web interface system also provides an instruction page where the various operators in our query language are explained and a number of examples shown (see Fig. 8). Users could click on the "Try it!" button to examine

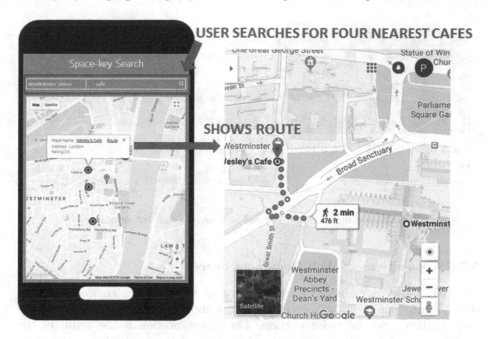

Fig. 9. Space-key search application for novice users[6].

the search results of the examples and could also freely modify the example operations.

Furthermore, another application of the map search which utilized our proposed query language (only using the "space-key" for a more simple and intuitive search) was also developed. This application was conceptualized by looking at how non-professional common users generally used location-based mapping services. Although route navigation was a commonly used feature, users also generally used location-based services to quickly identify different types of nearby venues and then find out how they could travel to such locations. Therefore, we developed an android application ("space-key search" application) which utilized the primitive unit of our spatial query language to allow users to search for nearby venues (users are able to search for nearby locations using only the space-key). For example, the user would enter the query "current location␣␣␣␣␣cafe", to find the four nearest cafes to them on auto adjust map zooming (see Fig. 9). Clicking on the markers would show details of the venue (the address, review scores etc.) as well as a link with the details of the route to the store. The application itself could be downloaded from Google Play store[5] (in Japanese). A mobile Web version of this application[6] was also developed for evaluation purposes.

---

[5] https://play.google.com/store/apps/details?id=com.kawaiLab.spatialQuery.
[6] https://yklab.kyoto-su.ac.jp/~sakata/simple/spatialQuary.

Fig. 10. Example of web document search (Color figure online).

5.2   Applications of Web Document Search

Figure 10 shows an example of a web document search by the keywords "Xiang Xiang" and "panda". The search results are shown in the yellow portion, the gray portion, and the green portion. The yellow portion is the result of the search query: Xiang Xiang␣^panda, which is used to search all the sentences between "Xiang Xiang" and "panda" where "panda" precedes "Xiang Xiang" (see Fig. 10(b)). The grey portion and the green portion are the results of the search query: panda␣^Xiang Xiang, which is used to search all the sentences between "panda" and "Xiang Xiang" where "Xiang Xiang" precedes "panda" (see Fig. 10(c)(d)). The green portion is the result of the query: panda␣#<4lines␣Xiang Xiang, which is used to search for all the sentences including the keyword "Xiang Xiang" are before 4 lines from the keyword "panda" (see Fig. 10(d)). The picture shown in the red color is the result of the search query: panda␣#<1line␣.wmv, which is used to search a .wmv file is before 1 line from the keyword "panda" (see Fig. 10(a)).

6   Preliminary User Study

To evaluate the potential usefulness of our proposed query language, a user evaluation study was carried out in which 15 students from a university-level computer science course were recruited and asked to carry out a series of location search tasks. Participants had previous experience in using Google Maps in their daily life and were not familiar with the places which were used in the search assignments. Overall, the main aim was to determine whether such an operation based query language would be feasible for developers to learn and use. Each participant was asked to use both our proposed query language through the Web interface system which was developed (spatial language condition) as well as through the Google Maps system (Google map condition) to complete 5 search assignments.

Each search assignment consisted of a task to search for places (e.g., pizza parlor) near a specific location (e.g., The White House). For example, one task consisted of trying to find the number of pizza parlors located within 500 m of Times Square. In another task, participants were asked to find the number of pizza parlors located within 700 m of Times Square which is also located 1000 m from the Empire State Building. Written instructions and examples were provided to help participants complete the tasks and introduce the various spatial operators. An objective measurement of performance was obtained by measuring the time participants spent on each task. To measure subjective user experience, the System Usability Scale (SUS) was used [2], which involved the rating of perceived effectiveness, efficiency, and satisfaction.

Overall, participants rated a higher SUS score for the spatial language condition (Mean = 70.17, SD = 13.09) than the Google map condition (Mean = 24.46, SD = 10.61) ($t(13) = 7.24$, $p < 0.001$)). In addition, participants were able to complete the tasks using less time (seconds) in the spatial language condition (Mean = 82.27, SD = 28.18) then the Google map condition (Mean = 180.86, SD = 49.25) ($t(13) = -8.219$, $p < 0.001$). Therefore, it seems that at least for search tasks which involve the combination and manipulation of spatial regions, the proposed query language search system could indeed be useful.

7 Conclusion

In this paper, we proposed a novel spatio-temporal query language for spatial and temporal search which can be used to express complex search queries in a text equation format. Also, we use "space characters" (the space-key) between keywords which are used to express the geographical distance or time-length between matching objects in an easy and intuitive way for general users. We implemented a prototype of our proposed system as a RESTFUL Web service and developed some applications of map search and web document search to showcase how the query language could be used. We also experimented with Google Maps for map search and the results from our user study highlighted the potential usefulness of our proposed query language.

In the future, we look to expand our spatio-temporal query language to other search domains such as image or video search. Although, we have shown how our proposed query language could be used in map search and web document search, our proposed query language could easily be applied to spatial and temporal search within images and videos as well. For example, a query searching for video frames within a video that contains the object "B" and are within 10 s from the object "A" is expressed by "A␣10sec␣B".

Acknowledgment. The work in this paper is partially supported by JSPS KAKENHI Grant Numbers 16H01722, 17K12686.

References

1. Spatial Databases with Application to GIS. Morgan Kaufmann Publishers Inc., San Francisco (2002)
2. Brooke, J.: SUS-a quick and dirty usability scale. Usability Eval. Ind. **189**(194), 4–7 (1996)
3. Duckham, M., Kulik, L.: "Simplest" paths: automated route selection for navigation. In: Kuhn, W., Worboys, M.F., Timpf, S. (eds.) COSIT 2003. LNCS, vol. 2825, pp. 169–185. Springer, Heidelberg (2003). https://doi.org/10.1007/978-3-540-39923-0_12
4. Elsevier: Elsevier R&D Solution: How can I search literature with reduced noise? Utilization of "proximity operator" in ScienceDirect & Scopus, 19 August 2015. http://jp.elsevier.com/_data/assets/pdf_file/0017/263240/Tips_Scopus_201508.pdf
5. Fu, K., Lu, Y.C., Lu, C.T.: Treads: a safe route recommender using social media mining and text summarization. In: Proceedings of the 22nd ACM SIGSPATIAL International Conference on Advances in Geographic Information Systems, pp. 557–560. ACM (2014)
6. Güting, R.H., Schneider, M.: Moving Objects Databases. Elsevier, Amsterdam (2005)
7. Hwang, E., Subrahmanian, V.: Querying video libraries. J. Vis. Commun. Image Represent. **7**(1), 44–60 (1996)
8. Khosla, R.: Oracle text reference: 11g release 2 (11.2), b61357, 06 October 2015. https://docs.oracle.com/cd/E16338_01/text.112/b61357/title.htm
9. Kim, J., Cha, M., Sandholm, T.: SocRoutes: safe routes based on tweet sentiments. In: Proceedings of the 23rd International Conference on World Wide Web, pp. 179–182. ACM (2014)
10. Liang, H.W., Hwang, Y.H.: Mobile phone use behaviors and postures on public transportation systems. PLoS ONE **11**(2), e0148419 (2016)
11. Lim, K.H., Chan, J., Leckie, C., Karunasekera, S.: Personalized tour recommendation based on user interests and points of interest visit durations. In: Proceedings of the 24th International Conference on Artificial Intelligence, IJCAI 2015, pp. 1778–1784. AAAI Press (2015). http://dl.acm.org/citation.cfm?id=2832415.2832496
12. Mata, F., et al.: A mobile information system based on crowd-sensed and official crime data for finding safe routes: a case study of Mexico City. Mob. Inf. Syst. **2016**, 1–11 (2016)
13. Pradhan, S., Sogo, T., Tajima, K., Tanaka, K.: A new algebraic approach to retrieve meaningful video intervals from fragmentarily indexed video shots. In: Arisawa, H., Catarci, T. (eds.) Advances in Visual Information Management. ITIFIP, vol. 40, pp. 11–30. Springer, Boston, MA (2000). https://doi.org/10.1007/978-0-387-35504-7_2
14. Pradhan, S., Tajima, K., Tanaka, K.: A query model to synthesize answer intervals from indexed video units. IEEE Trans. Knowl. Data Eng. **13**(5), 824–838 (2001)
15. Quercia, D., Aiello, L.M., Schifanella, R., Davies, A.: The digital life of walkable streets. In: International World Wide Web Conferences, pp. 875–884 (2015)
16. Quercia, D., Schifanella, R., Aiello, L.M.: The shortest path to happiness: recommending beautiful, quiet, and happy routes in the city. In: Proceedings of the 25th ACM Conference on Hypertext and Social Media, pp. 116–125. ACM (2014)
17. Su, H., Zheng, K., Huang, J., Jeung, H., Chen, L., Zhou, X.: CrowdPlanner: a crowd-based route recommendation system. In: 2014 IEEE 30th International Conference On Data Engineering (ICDE), pp. 1144–1155. IEEE (2014)

18. Sujeet, P., Tajima, K., Tanaka, K.: Interval glue operations and answer filtering for video data retrieval. IPSJ Trans. Databases (TOD) **40**(3), 80–90 (1999). https://ci.nii.ac.jp/naid/110002724879/en/
19. Wakamiya, S., Kawasaki, H., Kawai, Y., Jatowt, A., Aramaki, E., Akiyama, T.: Lets not stare at smartphones while walking: memorable route recommendation by detecting effective landmarks. In: Proceedings of the 2016 ACM International Joint Conference on Pervasive and Ubiquitous Computing, UbiComp 2016, pp. 1136–1146. ACM, New York (2016). https://doi.org/10.1145/2971648.2971758
20. Weiss, R., Duda, A., Gifford, D.K.: Composition and search with a video algebra. IEEE Multimed. **2**(1), 12–25 (1995)
21. Yin, H., Sun, Y., Cui, B., Hu, Z., Chen, L.: LCARS: a location-content-aware recommender system. In: Proceedings of the 19th ACM SIGKDD International Conference on Knowledge Discovery and Data Mining, pp. 221–229. ACM (2013)

Concept and Framework of 3D Geo-Spatial Grid System

Ji Song Yoo[1(✉)], Kyung Ju Min[1(✉)], and Jong Wook Ahn[2(✉)]

[1] Smart Urban Space Institute, Anyang University,
Anyang, Gyeonggi-do 14028, Republic of Korea
jisong14@gmail.com, kjgang@nate.com
[2] Department of Urban Information Engineering, Anyang University,
Anyang, Gyeonggi-do 14028, Republic of Korea
ajw0603@anyang.ac.kr

Abstract. As various types of information are being created through the recent technical developments in, and use of, smart devices, it is becoming increasingly important to manage spatial information, which is the foundation for utilizing information. This study defines the characteristics of spatial information, which can be used in the future, and proposes a 3D geospatial grid system (3DGGS) to manage spatial information. Based on a literature review, the characteristics of spatial information, which will be required in the future, are examined. Thus, the concept and characteristics of 3DGGS are presented, based upon which the application scope and rules of 3DGGS are set. In addition, geographic information systems developed by the Rep. of Korea and other countries are analyzed, and an architecture and framework through which 3DGGS can be implemented are presented. Finally, the results of this study are summarized, and areas of future works are introduced.

Keywords: 3D geo-spatial grid system · 3D geo-spatial · 3D grid system · 3D spatial visualization · 3D spatial analysis · 3D spatial data management

1 Introduction

The development of Information and Communication Technology (ICT) is accompanied by the creation of various types of information. Along with static spatial information, including records of geographic features, an increasing amount of dynamic real-time information is being produced with smart devices including autonomous vehicles and drones. In this regard, spatial information is becoming more important as the foundation for enhancing the intelligent aspects of our lives. Accordingly, it is becoming increasingly necessary to manage spatial information generated from various spaces and objects [1].

A type of spatial information computing called an early geographic information system (GIS) is utilized to organize layers of point, line, and polygon data using 2D coordinates on the ground. Layers were overlaid to find the spatial significance.

However, there are some areas in which it is difficult to apply positioning systems such as 2D coordinates or address systems. In such areas, particular locations cannot be

Y. Kawai et al. (Eds.): W2GIS 2019, LNCS 11474, pp. 136–149, 2019.
https://doi.org/10.1007/978-3-030-17246-6_11

distinguished or classified. For this reason, the data need to be accurately managed [2]. To deal with this issue, 2D grid systems have been constructed in which the space is plotted out for distinction and classification [3].

Next, remarkable 3D modeling technology for reproducing the real world has been developed to reduce distortions or the generalization of spatial information caused by a 2D representation of the real world, and to satisfy the enhanced demand for realism [4]. In other words, technology for representing the real world in cyberspace is being developed more quickly than ever before based on necessity [5]. The concept of 3D spatial information implies the role of a cyberinfrastructure enabling all areas of our daily lives to be continuously used in cyberspace [6]. Accordingly, different types of information on not only the ground surface and the facilities located on this surface, but also on aerial, underground, and underwater spatial locations, need to connecting based on the continuous and constant criterion. Such spatial continuity improves the reality of spatial information and enables continuous spatial information to be provided [7].

Spatial information will need difference expression ways for spatial information of being connect because expanded from recognizing area by human sense (eyes) to unrecognized area. Ours senses can recognize connectivity of air, underground, water space and land and saves this information. This ability is called this spatial perception.

Future device in 4th industrial revolution would need this ability like human. Autonomous device is need to recognizing space. Drone is flying sky we can't see. They are using positioning system based on lands, GPS or GNSS supporting. These are relative reference systems. This environment make drone depends on other system. Therefore they need to absolute reference system for get completely autonomous flying themselves. Based on this idea we suggest to spatial information management system different now.

This study proposes a 3D geo-spatial grid system (3DGGS) which is a management system of 3D spatial information, and a criterion for reflecting spatial information of the real world without a loss in continuity or reality. Based on a literature review, the concept, characteristics, role, and functions of 3DGGS are defined and its framework is presented.

2 Literature Review

In this section, 3DGGS is defined and its characteristics are identified through a review of existing studies on the role of spatial information in the future world and 3D grid systems. In the modern world, spatial information role is a cyber infrastructure that can react to the changes and technical development for the future, as well as reflect the fusion of information in various spaces [8].

The advent of the 4th industrial revolution is one of the biggest issues for our future. People are extremely likely to face an overall change to their living environment, including their employment. At the societal level, the digital revolution will open up a new era of technological convergence where the boundaries of physical, digital, and biological areas will become blurred [9].

Automation and artificial intelligence will also change the type of economic development that occurs, and new data processing methods for storing, managing, analyzing, and utilizing fused information will emerge [10]. New data processing methods will enable advanced technology to be freely utilized in everyday life by removing spatial restrictions and obscuring the boundary between reality and virtual reality [11]. In this regard, our future will have an obscure boundary between spaces that have been distinctively defined. In particular, the future will be greatly affected by IoT, big data, artificial intelligence, and spatial developments, all of which will be applied across spatial boundaries [12].

According to a study on a positioning reference system for the hyper-connective society in the Rep. of Korea, such a positioning reference system needs to be dynamically implemented for everyday life in the future. A seamless and dynamic positioning reference system with about a 1-m scale will be required to operate across ground, underground, aerial, and indoor spaces [4]. In addition, as IoT and sensor technologies are developed, spatial information will be more actively utilized, and a real-time integrated management system will be developed in the future [13].

Future technical and social trends, which have been reviewed thus far, are summarized in Table 1. Spatial continuity (including aerial, ground, underground, and underwater spaces), the up-to-date status of real-time information collected by sensors, the positioning accuracy for any spaces (indoor and outdoor), and the convergence of information are key issues with regard to future spatial information. Accordingly, 3DGGS should function as a reference or hub of future spatial information, where enormous amount of data can be connected without spatial constraints.

Table 1. Literature review: global future technology trend.

Area	Institute	Key issue
World Wide	WEF [9]	A new era of technology convergence will emerge in which physical, digital, and biological spaces will be seamlessly connected
	NIC (2017) [10]	Human life will undergo unprecedented alterations owing to changes in the global environment, which is attributable to climate and social changes caused by automation and artificial intelligence
	Gartner [11]	IoT, intelligent, and conversational platforms will emerge. These platforms will mediate reality and virtual reality
South Korea	MSI [12]	Fifteen key technologies including in the areas of IoT, big data, and AI will influence the future
	KASM [4]	The hyper-connected future society will obscure the boundary between virtual and physical worlds. This new world will mix people, spaces, information, and smart devices
	Park [13]	IoT and sensor technologies using spatial information are expected to enable an integrated survey of environmental information, social conditions, and changes to ground features, and to bring about systems for monitoring, collecting, fusing, analyzing, and managing data in real time

Table 2 lists 3D spatial information technologies. Because 3D spatial information can be used to visualize the real world quite closely, such information can be more effectively identified and easier to utilize than 2D spatial information [14]. However, there have been difficulties in the geometric computations and calculations of the relationships between objects. In addition, models have to be selectively used [15].

As computing technology has gradually developed, some researchers have predicted that property data will be directly contained in 3D cubic type geographic information [16]. It has also been anticipated that things or users in motion will use accurate spatial information in real time [5]. Currently, beyond the traditional GIS field, 3D spatial information is being utilized to visualize analysis or evaluation results in various areas of research [17].

Table 2. Literature review: needs for 3D spatial information technology.

Area	Author	Key issue
World Wide	Tavaniti et al. [14]	Three-dimensional spatial information is more effectively perceived by users, and is easier to use than 2D spatial information
	Zlatanova et al. [15]	For a 3D spatial analysis, models need to be used appropriately. For example, geometric models can be used to calculate locations, whereas relationships can be calculated based on the topology
	Berry [16]	Three-dimensional spatial information will evolve to contain spatial information in a visualized topography
	Biljecki et al. [17]	Most 3D urban models are utilized for visualization, and such applications tend to extend beyond the traditional GIS area
	Liang et al. [18]	Although 2D models using planes are quick and easy for certain tasks, 3D models can calculate and visualize the surrounding areas of a building and the effect of an envelope
South Korea	Kang et al. [5]	Spatial information in the future should ensure the mobility of geographical information
	Seo et al. [1]	Three-dimensional spatial information needs to have completeness, logical consistency, locational accuracy, punctuality, and subject accuracy
	Park et al. [19]	As different ranges of a virtual space are expressed depending on the basic elements of the 3D data model, it is necessary to define an appropriate data model and basic elements for information or analyses that will be used for a virtual space
	Cho et al. [20]	A fused spatial information service demanded for a smart society should realize mobility, customizability, and real-time support
	Park [13]	High-precision spatial information needs to be developed to reduce errors, save resources, and achieve energy efficiency in autonomous vehicles

Existing 2D spatial information provides a quick analysis for planes, whereas 3D spatial information provides accurate insight by enabling an analysis to reflect various three-dimensional factors [18]. Three-dimensional spatial information is particularly essential to unmanned smart devices that will be widely used in the future because such devices will freely move in space. Based on an analysis of accurate information on different space areas and locations, a judgment needs to be made in advance [13]. The early grid system used longitude and latitude lines forming a grid with regular patterns on the earth's surface [21].

Table 3. Literature review: trend in grid system use.

Area	Author	Key issue
World Wide	Goodchild [21]	A traditional grid system applies a 2D latitude and longitude grid of the earth's surface
	Sahr [2]	Global grid models can be distinguished into tetrahedral, octahedral, dodecahedral, and icosahedral models according to the projection and coding methods
	Mahdavi-Amiri et al. [3]	DGGS is used to index spatial information of the earth and allocate data in the Digital Earth Framework
South Korea	Yang [22]	Residential areas can be positioned using address systems, whereas fields and mountains are difficult to position with existing systems. To solve this problem, the National Point Number consisting of 2D grids was introduced
	Kim [23]	Data from all areas including the oceans, weather systems, and land are integrated and standardized in grid form. Thus, the accuracy of the data is sufficiently improved to realize interoperability

This model has various types depending on the projection and coding methods applied [2]. Recently, the Digital Earth Framework utilized Discrete Global Grid Systems (DGGS) to index every space on the ground surface and allocate data to it [3]. In particular, DGSS is used to identify locations in fields, mountains, and sea surface areas, where positioning has been difficult to achieve, and thus spatial information becomes more applicable [22]. Such grid systems are often utilized in different ways in diverse areas such as the oceans, weather, and land. Because different applications are integrated and standardized in grid form, data can be provided with sufficient accuracy to allow them to be made available across different areas [23]. Table 3 shows the trend in the use of grid systems.

3   Definition of 3DGGS Characters and Concept

3.1   Establishment of Concepts and Characteristics

As is clear from the above literature review, people, unmanned aerial vehicles, autonomous cars, and other subjects with high mobility in a smart society can effectively use spatial information only when the limitations of the existing positional reference systems are improved [5]. Accordingly, to express all spaces (aerial, ground, underground, and underwater) as perceived by humans, 3DGGS needs to index and visualize those spaces as well as allocate and manage the data for each space. In this regard, 3DGGS should be up-to-date, accurate, and continuous. These features enable flexible control by users with regard to the addition or modification of data [24]. Consequently, the concept of 3DGGS can be set as a spatial information management system functioning as a single positional reference system for all spaces. Figure 1 help to intuitionally understand concepts 3DGGS.

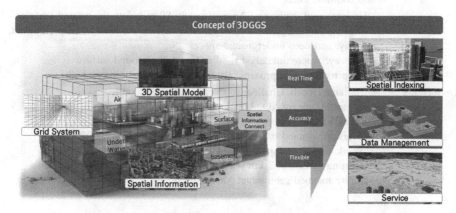

Fig. 1. Concept of 3DGGS

3.2   Establishment of Concepts and Characteristics

The implementation scope and criteria of 3DGGS need to be defined in order to utilize the system as a single positional reference. Because the majority of current spatial information is based on transverse Mercator coordinates, a spatial partition needs to adopt the longitudes/latitudes for storage and integrated data management. Spherical and spheroid grids are available for spatial partitions. A spherical grid does not consider the height, whereas a spheroid grid does not consider the ground surface data [25].

3DGGS adopts the longitude/latitude coordinate system for grid partitions but uses a spheroid grid to consider the height (depth). For this reason, the range of height (depth) should be set. The height and depth can be set based on the scope of human activity, as shown in Table 4. As for the height, the range can be approximately 10 km from the ground surface by reflecting weather phenomena or airplanes flying in the air. With regard to the depth, when including underground facilities and underwater spaces in which manned or unmanned submarines can travel, the Mariana Trench is the deepest point, at approximately 11 km. Accordingly, because 3DGGS aims to manage spatial information generated through the scope of human activity in three dimensions, the partitioning range of a spherical grid is set to 10 km in height and 10 km in depth.

If an ID is assigned to a partitioned grid, as shown in Fig. 2, the bottom-left portion of the map will have the index (0, 0) for the rows and columns, respectively. The height ID is given separately. The value of the row is zero at the bottom and increases as it moves upward. The column has a value of zero at the leftmost position, and increases as it moves to the right.

Table 4. Height regarding the scope of human activity.

Space	Scope
Air	- Weather phenomena: 10 km or less from the ground surface - Airplane operation: 10 km from the ground surface
Underground	- Underground facilities: Approximately 20 m below the ground surface * Underground facilities: Water supply, sewage, communication, gas, heating, and common ducts
	- Underground structures: Approximately 80 m below the ground surface * Underground structures: Passage, roadway, commercial zone, parking lot, subway, and deep underground spaces • Subway: Approximately 13–80 m below the ground surface • Deep roadway: Approximately 40–60 m below the ground surface
Underwater	- Mariana Trench, Challenger Deep: 11,034 m deep - Domestic unmanned submarine: Maximum diving depth of 6 km * HEMIRE: 6 km deep - Overseas manned submarines: Maximum diving depth of 7 km * Shinkai 6500 of Japan: 6.5 km deep * Jaorung of China: 7 km deep - Future manned submarine: Maximum diving depth of 12 km

The 3D grid of 3DGGS can be divided into a 3D tile utilizing Tile Map Service and multiple numbers of 3D grids based on the tile. In general, a grid system partitions data from higher to lower levels. As presented in Table 5, the 3D tile sets the size of the highest grid to 1° of longitude (approximately 32.421 km) at 36° latitude, which includes the Rep. of Korea, in the longitude/latitude coordinates.

Table 5. Level of cube.

	Distance in Longitude (m)		
Latitude	35	36	37
Level 0	3282753.18	3242136.95	3200533.14
Level 1	1641376.59	1621068.48	1600266.57
Level 2	820688.30	810534.24	800133.29
...	...	...	...
Level 18	12.52	12.37	12.21
Level 19	6.26	6.18	6.10
Level 20	3.13	3.09	3.05

Twenty lower levels are obtained by dividing the longitude and latitude lines of the highest grid in half 20 times. Thus, useful 3D tiles with a size of approximately 3.09 m can be organized. It is important for the 3D tiles to divide a space equally. Accordingly, even if a single grid is used, the association with the lower grids needs to be considered, and a basic system for internal data management is to be provided [25].

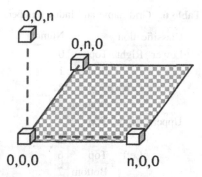

Fig. 2. 3DGGS indexing of x (long), y (lat), z (height)

The height ID (number) is allocated in a different way. Heights have previously been considered simple values; however, it is now necessary to define the heights of the grid. Because the longitude/latitude coordinate system does not include any reference to height, to assign a height, each angle is converted based on the circumference at the equator. When the height IDs are numbered when assigning zero to sea level or a geoid, the IDs for the underground points, or those points below sea level, have negative values. If the IDs are not numbered with negative values, they can be managed more efficiently. Accordingly, as shown in Fig. 3, the deepest point in the vertical direction has a value of zero, and positive numbers are added to ensure the consistency of the ID system [28].

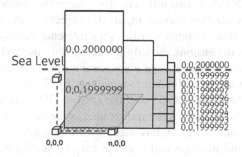

Fig. 3. Numbering height for 3DGGS.

Figure 4 and Table 6 shows the naming convention used for expressing the relationship between the parent and child, which prevents any confusion with regard to the data connection and clarifying the data association. If the ID of a grid indicates an absolute position, this naming convention shows a relative position when a parent manages a child. Basically, the hierarchy is organized.

Table 6. Grid name and index number.

Classification			Number
Lower	Right	Top	0
		Bottom	1
	Left	Top	2
		Bottom	3
Upper	Right	Top	4
		Bottom	5
	Left	Top	6
		Bottom	7

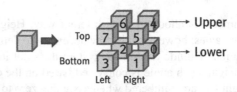

Fig. 4. Illustration of grid name and index number.

3.3   Core Functions of 3DGGS

3DGGS not only visualizes spatial information but also manages data by partitioning the data into uniform grids. In this regard, the data management is expanded from a 2D grid to a 3D grid. 3DGGS is also different from currently available 3D visualization services. The 3D visualization service inputs 3D objects in the x and y coordinates based on a 2D grid system. In other words, every reference exists on the 2D plane and is visualized in three dimensions. Strictly speaking, this method is not a data management system but is quite similar to a visualization model. Accordingly, 3D analysis and data management are more difficult to achieve [26]. Table 7 shows the differences in data management between 3DGGS and the existing 3D modeling.

The existing 3D visualization system uses tiles as the unit of data management. As illustrated on the left side of Fig. 5, this method divides a given surface into 2D grids with a uniform size and manages and displays data by storing and nesting them into each grid on a required object basis. However, data redundancy does occur. For this reason, as shown in Fig. 5, to completely contain the information using only a single file without additional external information, 3DGGS partitions and manages the data within grids.

In addition, the 3D visualization system visualizes grid data by reading and expressing a model at an appropriate position. 3DGGS uses voxel-based grid data and thus a GPU can be fully utilized. When a GPU is utilized, a greater volume of data can be calculated than with an existing method. Consequently, significant improvements in both speed and convenience are expected [27].

Table 7. Differences in data management.

Classification	3D modeling	3DGGS
Basic	Data management in 2D tile partitions with no height being considered	All 3D information of target objects is partitioned in a grid system for management
2D Raster	Transformation and visualization on three-dimensional surface through DEM	Transformation into three-dimensional surface through DEM and visualization of each grid containing data
2D Vector	Modeling and insertion into a reference point as a model	3D grid data management
3D Raster	Tile-based images are visualized on a topography made using DEM	Tile-based images are visualized on a topography made using DEM
3D Vector	Models are inserted into each input point designated based on the tile	Data are partitioned and managed based on grid
Empty Space Management	Not supported	Information of vacant spaces is also allocated to a grid and managed as a single unit
Raster Visualization	Model creation at CPU and visualization	Raster data are transferred to GPU and then visualized
Vector Visualization	Object-based visualization	Grid-based visualization
Data Search	Data search at 2D scale and along the altitudes	Data search at 3D scale

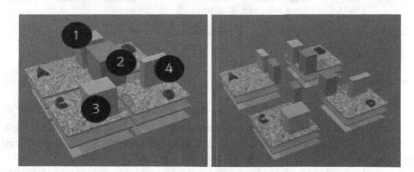

Fig. 5. Data partitioning and storage. Object 2 (red) contains information on all four tiles (A, B, C, and D), and non-redundancy can be set through visualization. However, data redundancy does occur. When Grids A, B, C, and D are loaded, all object information is stored together. For this reason, object redundancy does not occur because the partitioning and information can be utilized using only a single grid. (Color figure online)

4 3D Geo-Spatial Grid Framework

4.1 Concepts of 3DGGS Framework

As previously pointed out, 3DGGS needs to be up-to-date, accurate and continuous. This section describes a case study presenting the 3DGGS framework. As shown in Table 8, spatial information systems used in the Rep. of Korea and other countries have been examined.

Table 8. Spatial information systems and their functions.

Classification	QGIS	GRASS	SAGA	V-World	GSBD	GEEPS
Development approach	Open source	Open source	Open source	Government plan	Government plan	Open source
Programming language	C++, Python, Qt	C, C++, Python, Tcl	C, Python, Java	Java, HTML, CSS, JavaScript, etc.	Java, HTML, CSS, JavaScript, etc.	C, C++, Python
Data conversion/processing	GDAL/OGR	**GRASS Row Input/Output**, GDAL/OGR	SAGA GIS, GDAL/OGR, etc.	**V-World GIS**	**GSBD ETL**	GDAL/OGR
Data storage/management	PostgreSQL, Post GIS, MS SQL Server	**GRASS Database Management Interface**	PostgreSQL, PostGIS, etc.	PostgreSQL, PostGIS, Altibase	NoSQL, Hadoop	Mysql, PostGIS
Data analysis	QGIS Core, LAStools, OTB, SAGA, TauDEM	**GRASS Array Statistics, GRASS Raster, GRASS 3D Raster Volume, etc.**	SAGA GIS, Cellular AUtomata, Erosion, IHACRES, RivFlow, etc. (about 700 programs)	anyGuide Map, **XDWorld**, Mariner	**GSBD analysis modules**	u-GIS, flex, bison, CMake
Data visualization	QGIS Core, LAStools, OTB, SAGA, TauDEM	**GRASS Display, GRASS Cairo Display**, Graphics monitor driver, etc.	SAGA GIS, 3D Viewer, Grids, Point Clouds Viewer	anyGuide Map, **XDWorld**, Mariner	**GSBD Visualization modules**	u-GIS, flex, bison, CMake
User interface	Qt, **QGIS GUI**	**GRASS Display, GRASS Cairo Display**, Graphics monitor driver, etc.	SAGA GIS GUI, etc.	**V-World GUI**	GSBD GUI	Qt, Pyqt

*Items marked in bold type were developed using GIS.

A conceptual model of 3DGGS was constructed based on the 3DGGS characteristics and the configuration of the spatial information system, which were identified through a literature review, as shown in Fig. 6.

Figure 7 shows the framework constructed using the conceptual model. 3DGGS consists of an information area, a core area, and a service area. The information area includes the 3DGGS data ETL (Extract, Transform, Load) layer and data management layer linking various types of information, and thus responds to the fusion and convergence of spatial information. Data in the information area are transformed into a 3D geo-spatial grid, which is the format used in 3DGGS. This basic process for storage and management enables a 3D grid-based analysis. The core area provides a visualization engine and an analytical engine for both 2D and 3D raster and vector analyses. In particular, this area is the key aspect of 3DGGS, which enables a spatial operation

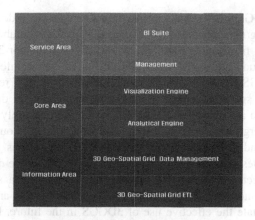

Fig. 6. 3DGGS conceptual model.

based on a 3D grid. Finally, the service area manages 3DGGS and provides data for developing applications. This area consists of a management area and BI suite (Business Intelligence suite).

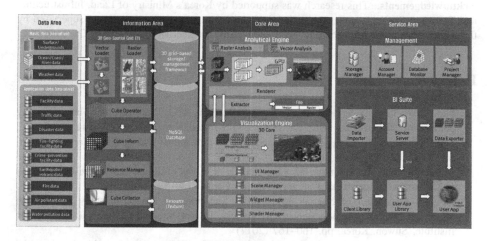

Fig. 7. 3DGGS framework.

5 Conclusions and Future Work

The present study aimed to establish the concept and framework of a 3D geo-spatial grid system (3DGGS). With the development of ICT, future spatial information needs to be fused across all types of spaces by ensuring such characteristics as continuity between spaces, up-to-date real-time information capability, and positioning accuracy. Accordingly, it is becoming necessary to construct spatial information using a single positional reference such as 3DGGS.

In particular, 3DGGS is basically a system that enables 3D indexing, visualization, and data partitioning/management for all spaces. To reflect the above characteristics, which were derived from a literature review and are required for 3DGGS, this study conducted a case study and constructed 3DGGS based on the information, core, and service areas. 3DGGS fuses various data and put information of real worlds, it make between which the space boundary becomes for connection in digital space. Thus, various types of data are placed into 3D grids, which are uniformly distinguished, and aerial, ground, underground, and underwater data can be continuously processed. The core area analyses and visualizes the real world in three dimensions. Finally, the service area provides modules through which application programs can be developed based on 3DGGS. It is projected that 3DGGS will be widely used as a spatial information technology for the integrated management of data across diverse areas. Future studies are needed to facilitate the effective use of 3DGGS in the future. Because the height reference presented in this study is based on the longitude/latitude coordinate system, an institutional agreement is required for international use.

In this regard, standardization needs to be continuously discussed with experts from relevant areas. Moreover, it will be necessary to discover new services that can be made available based on 3DGGS, and propose them as reference models.

Acknowledgements. This research was supported by Korea's Ministry of Land, Infrastructure, and Transport (MOLIT) under the national spatial information research program supervised by the Korea Agency for Infrastructure Technology Advancement (KAIA, 18NSIP-C135506-02).

References

1. Seo, C.W., Choi, Y.S., Kim, J.M., Kim, Y.H., Kim, Y.G.: Improvement plan of quality control for 3D geospatial database. Proc. J. GIS Assoc. Korea **17**(2), 231–241 (2009)
2. Sahr, K., White, D., Kimerling, A.J.: Geodesic discrete global grid systems. Cartogr. Geogr. Inf. Sci. **30**(2), 121–134 (2003)
3. Mahdavi-Amiri, A., Alderson, T., Samavati, F.: A survey of digital earth representation and visualization. University of Calgary, Calgary, Canada (2015)
4. Korean Association of Spatial Information: Surveying & Mapping: study on position reference system for hyper-connectivity society. Report National Geographic Information Institute, Suwon, Korea, pp. 160–167 (2017)
5. Kang, H.K., Lee, S.Z., Li, K.J: Position identification scheme for ubiquitous spatial computing. In: Proceedings of 2008 Korea GIS Society Symposium, Korea, pp. 280–285. Korea Spatial Information Society (2008)
6. Wright, D.J., Wang, S.: The emergence of spatial cyberinfrastructure. PNAS J. Nat. Acad. Sci. **108**(14), 5488–5491 (2011). https://doi.org/10.1073/pnas.1103051108
7. Wang, S., Wilkins-Diehr, N.R., Nyerges, T.L.: CyberGIS—Toward synergistic advancement of cyberinfrastructure and GIScience: a workshop summary. J. Spat. Inf. Sci. **4**, 125–148 (2012). https://doi.org/10.5311/JOSIS.2012.4.83
8. Dawn, J.: Wright: theory and application in a post-GISystems world. Int. J. Geogr. Inf. Sci. **26**(12), 2197–2209 (2012)
9. World Economic Forum: The Future of Jobs. Report World Economic Forum, Cologny, Switzerland (2016)

10. National Intelligence Council: Global Trends 2030: Alternative Worlds. Report National Intelligence Council (2012). ISBN: 978-1-929667-21-5
11. Gartner: Technology Trends for 2018, Report Gartner, Published No. G00327329, Stamford, USA (2017)
12. Ministry of Science and ICT: Korea Future Issues 2015 (Eng Version). Report Future Preparatory Committee, Gwacheon, Republic of Korea (2015)
13. Park, M.S.: 2017 Bring happiness to the land. Korea Land and Geospatial Informatix Corporation, Jeonju, Korea, pp. 22–39 (2017)
14. Tavanti, M., Lind, M.: 2D vs. 3D, implications on spatial memory. In: Proceedings of IEEE Symposium on Information Visualization, San Diego, USA, pp. 131–138 (2001)
15. Zlatanova, S., Abdul Rahman, A., Shi, W.: Topology for 3D spatial objects. Comput. Geosci. **30**(4), 419–428 (2003). https://doi.org/10.1016/j.cageo.2003.06.004
16. Berry, J.K.: Map analysis: understanding spatial patterns and relationships. Geo Tec Media (2007). ISBN: 978-0974861319
17. Biljecki, F., Stoter, J., Ledoux, H., Zlatanova, S., Çöltekin, A.: Applications of 3D city models: state of the art review. ISPRS Int. J. Geo-Inf. **4**, 2842–2889 (2015). https://doi.org/10.3390/ijgi4042842
18. Liang, J., Gong, J.: A sparse voxel octree-based framework for computing solar radiation using 3D city models. ISPRS Int. J. Geo-Inf. **6**(4), 106–120 (2017). https://doi.org/10.3390/ijgi6040106
19. Park, S., Lee, J.: Comparative analysis of 3D spatial data models. J. Korea Spat. Inf. Soc. **17**(3), 277–285 (2009)
20. Cho, C.M., Hwang, N.S.: Realization strategy of convergence GIS based on spatial hierarchy. J. Korea Spat. Inf. Soc. **21**(5), 83–91 (2013)
21. Goodchild, M.F.: Discrete global grids for digital earth. In: Proceedings of International Conference on Discrete Global Grids, Santa Barbara, USA, pp. 1–9 (2000)
22. Yang, S.C.: A study on the introduction of the national point number for advanced location-finding. J. Korean Soc. Surv. Geodesy, Photogramm. Cartogr. **31**(2), 151–157 (2013). https://doi.org/10.7848/ksgpc.2013.31.2.151
23. Kim, D.H., Kim, J.M., Yoon, B.C., Chang, E.M., Choi, Y.S.: Development plan of grid system utilizing spatial information. J. Korea Spat. Inf. Soc. **23**(6), 43–55 (2015)
24. Yoo, J.S., Min, K.J., Kim, M.K., Ahn, J.W.: A study on the necessity and direction of 3D grid system development. In: Proceedings of the Korean Society for Geo-Spatial Information System, Jeonju, Korea, pp. 236–239 (2017)
25. Wan, G., Cao, X., Li, F., Li, K.: Sphere shell space 3D grid. In: Proceedings of ISPRS WebMGS 2013 & DMGIS 2013, Xuzhou, Jiangsu, China, pp. 77–82 (2013)
26. Yoo, J.S., Min, K.J., Kim, M.K., Ahn, J.W.: A study on utilization based on 3D grid system. In: Proceedings of Korea Spatial Information Society, Jeju, Korea, pp. 152–156 (2018)
27. Laine, S., Karras, T.: Efficient sparse voxel octrees. In: Proceedings of ACM SIGGRAPH Symposium on Interactive 3D Graphics and Games (I3D), New York, USA, pp. 55–63 (2010). https://doi.org/10.1145/1730804.1730814
28. Lipman, Y., Kopf, J., Cohen-Or, D., Levin, D.: GPU-assisted positive mean value coordinates for mesh deformations. In: 2007 Proceedings of the Fifth Eurographics Symposium on Geometry Processing, Barcelona, Spain, pp. 117–123 (2007)

A Proposal of Distributed Occupancy Grid Map on Block Chain Network

Yousuke Watanabe(✉)

Institutes of Innovation for Future Society, Nagoya University,
Furo-cho, Chikusa-ku, Nagoya-shi, Aichi-ken 464-8601, Japan
watanabe@coi.nagoya-u.ac.jp

Abstract. In recent years, mobile robots such as autonomous vehicles and drones are widely tested in public spaces. A mobile robot uses embedded sensors to detect whether an area is occupied or vacant. Occupancy grid map is a popular data representation for handling a set of area states (occupied, vacant, unknown). To shrink unknown areas not covered by any sensors, sharing sensor data from multiple mobile robots becomes important. However, centralized systems are not suitable for information sharing in public spaces, because initial cost and operating cost are too expensive for one company. To avoid the problem, this paper proposes a decentralized information sharing system based on block chain network. Distributed ledgers on block chain network is used to maintain occupancy grid map in public spaces. To increase incentive of information provision from mobile robots, the proposed system exchanges area states reported from mobile robots for virtual currency. A mobile robot can receive pay when it uploads an area state, and it can buy the latest state of another area reported by other robots. This paper also includes the preliminary experiment of the proposed system.

Keywords: Occupancy grid map · Block chain · Spatial data management

1 Introduction

Mobile robots are popularly used in our life. Robot-vacuums move around in homes and offices, and robot-wagons carry containers in factories and warehouses. A mobile robot uses embedded sensors to detect whether an area is occupied or vacant. It dynamically selects an available route to execute tasks assigned to it. Occupancy grid map [1] is a popular data representation for handling a set of area states (occupied, vacant, unknown).

Mobile robots have been operated in restricted and isolated fields. Furthermore, they are tested in public spaces toward the practical use. An autonomous vehicle is an example of mobile robots in public spaces. It has high-resolution sensors such as LiDARs and cameras (Fig. 1). The vehicle recognizes surrounding environment by matching sensor data and high-definition road maps, and it

Y. Kawai et al. (Eds.): W2GIS 2019, LNCS 11474, pp. 150–159, 2019.
https://doi.org/10.1007/978-3-030-17246-6_12

Fig. 1. An example of mobile robots (autonomous vehicle)

autonomously performs maneuver control. Field-Operation-Test of autonomous vehicles have been started in several counties.

A drone is another example of mobile robots in public spaces. Its flight is completely controlled by a program. Environmental monitoring and baggage delivery are expected applications of drones. Since low altitude on private lands is restricted, drones need to flight on public spaces. As described above, mobile robots will be operated in not only private fields, but also public spaces.

A range of embedded sensors on each mobile robot is finite. Blind angles caused by obstacles are unavoidable. For example, a range of LiDAR in autonomous vehicles is about 100 m. And, laser light from LiDAR is easily blocked by obstacles. To shrink unknown areas not covered by any sensors, sharing sensor data from multiple mobile robots becomes important.

In restricted and isolated operation fields such as offices and factories, one company can manage all mobile robots. It is easy to develop a centralized system for information sharing in these fields. However, in public spaces, centralized systems are not suitable for information sharing. Because, mobile robots in public spaces are managed by multiple companies, so there is no leader of these companies. And, initial cost and operating cost of database servers and communication services are too expensive for one company.

To share sensor data in public spaces, this paper proposes a decentralized information sharing system based on block chain network. Block chain is a distributed data management technology for recording exchange of virtual currency such as Bitcoin [3]. We make block chain maintain occupancy grid map. Grid data are distributed among participants in block chain network, thus there is no centralized administrator. To increase incentive of information provision from mobile robots, the proposed system exchanges area states reported from mobile robots for virtual currency. In this research, we develop a prototype system by using open source block chain implementation, and we show our experimental result of the system in a private block chain network.

The remaining part of this paper is organized as follows: Sect. 2 explains occupancy grid map and block chain network. Section 3 presents the proposed system. Section 4 describes an implementation of the proposed system. Section 5 shows the preliminary experiment using the prototype system. Finally, Sect. 6 concludes the paper and mentions future research issues.

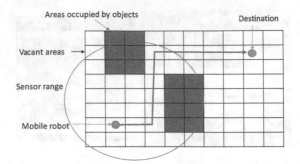

Fig. 2. Occupancy grid map

2   Preliminaries and Assumptions

2.1   Occupancy Grid Map

In the research area of mobile robots, occupancy grid map [1] is a popular data representation for handling area states. A field is divided into a two dimensional grid consisting of multiple cells (Fig. 2). Each cell stores the current state of the corresponding area, and it expresses whether the area is occupied or vacant. There are three state types: "occupied", "vacant", "unknown". "Occupied" means that there are some conflicting objects in the area. "Vacant" means that there is no object in the area. "Unknown" means that the area is out of the sensor range. ROS (Robot Operating System), which is one of famous middleware in mobile robot research, also provides a library to handle occupancy grid map [2].

In this paper, we assume that all mobile robots have sensors and communication devices. Each mobile robot can scan cells within its sensor range. In addition to that, it wants to know about further cells to choose an available route to the destination.

2.2   Block Chain Network

Block chain is a distributed data management technology for recording exchange of virtual currency such as Bitcoin [3] and Ethereum [4]. A participant node in a block chain network manages ledgers composed of block chain. Since ledger is a distributed storage in the network, we do not need centralized servers and administrators. Here, we provide just a summary of block chain, because details is out scope of this paper.

Virtual Currency. Virtual currency like Bitcoin uses public key cryptosystem to save owner information. Each coin is assigned to owner. A coin is encrypted by owner's public key, so owner can decrypt the coin by using his private key.

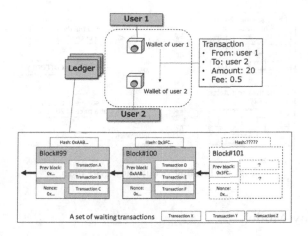

Fig. 3. Block chain

Suppose User A want to transfer his coin to User B. Firstly, User A decrypts the coin by using A's private key. Secondly, User A encrypts the coin by using B's public key. After that, the coin is encrypted by B's public key, thus the owner of the coin becomes B.

Transactions and Blocks. A basic unit of money transfer is called a transaction. A transaction contains old owner (from), new owner (to), amount, fee and so on (Fig. 3). A distributed ledger consists of a chain of blocks, and each block stores multiple transactions. A block has a sequential number and a unique hash value. The hash value of the previous block is stored in the next block, thus the next block refers to the previous block. Referred blocks makes a chain of blocks.

When a participant node in block chain network creates a new block, the node must follow the rule that a hash code of new block should be less than the threshold. Therefore, the node needs to find an appropriate combination of transactions and a Nonce value to be stored in the block. If the node successfully creates a new block, it can receive transaction fees in the block.

In this research, our prototype system is developed on Ethereum, which is one of open source implementations of block chains.

3 Proposed Method

As described above, we employ block chain for sharing occupancy grid map in public spaces. Sensor data detected by mobile robots are exchanged for virtual currency in the system (Fig. 4). A mobile robot can receive pay when it uploads an area state, and it can buy the latest state of another area. Providing a new state of each cell has the same value as querying a new state of each cell. And, participant nodes of block chain network also receive pay for maintaining ledgers.

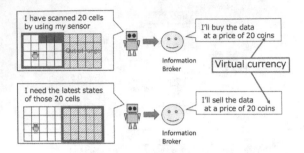

Fig. 4. Trading occupancy grid map

3.1 Economy Model

Here, we explain data objects exchanged in our system.

- **Cell state** $c_{x,y,t}$ F We consider three-dimensional grid: two spatial axes (x and y) and one time-axis (t). The occupancy grid map consists of three-dimensional cells. Each cell has a state, which may be "occupied" or "vacant" or "unknown". A cell state is assigned to owner.
- **Ticket for querying cell state** qF When a mobile robot requests the latest cell state, a ticket for querying is needed. A ticket q has the same value as a state of each cell $c_{x,y,t}$.

3.2 Participants

Figure 5 shows actors in our system. A participant node in block chain network can play one or more actors in the figure.

- **Ledger manager:** Many and unspecified ledger managers record transactions of exchanging cell states and tickets. Ledger manager receives a fee for transaction processing.
- **State publisher:** State publisher reports cell states observed by sensors. When State publisher sells a cell states to State broker, it receives a ticket for querying.
- **State broker:** State broker buys a cell state from State publisher, and it sells the cell state to State subscriber.
- **State subscriber:** State subscriber sends a ticket to buy a required cell state.

Generally, mobile robots act both State publisher and State subscriber. They can sell cell states which are covered within a range of embedded sensors. And, they can buy cell states which are out of their sensor ranges.

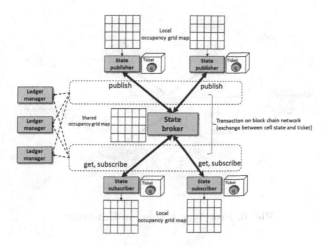

Fig. 5. The proposed framework

3.3   Data Exchange

We show data exchange flows in the proposed system. We define three operations: Publish, Subscribe and Get.

Publish Operation. Publish is an operation when State publisher observes the latest cell state by using sensors and provides the sensor data.

1. State publisher p obtains $c_{x,y,t}$, which is the cell state of coordinate (x, y) at time t, by using its embedded sensors.
2. State publisher p poses a trade of cell state $c_{x,y,t}$ to State broker b.
3. State broker b checks whether the cell state $c_{x,y,t}$ has been already reported. If not reported yet, b sends a ticket for querying p as payback to p. If $c_{x,y,t}$ exists, b does not send payback. But the reported data is used to validate existing $c_{x,y,t}$ (the validation method is included in future research issues).
4. In case of payback, State broker b becomes owner of $c_{x,y,t}$.
5. A fee of transaction processing is given to Ledger manager a.

Subscribe Operation. Subscribe is an operation when State subscriber declares required cell states and receives them.

1. State subscriber u sends a requirement (subscription) for cell (x, y) to State broker u.
2. When State publisher p provides the latest cell state $c_{x,y,t}$, State broker b notifies u of data arrival.
3. State subscriber u decides whether he actually buys the notified cell state $c_{x,y,t}$. If u needs the cell state, u sends a ticket for querying to State broker b.

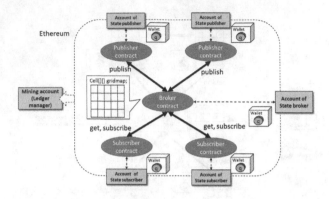

Fig. 6. Prototype system on Ethereum

4. State broker b returns the cell state $c_{x,y,t}$ to State subscriber u. b continues the owner of $c_{x,y,t}$.
5. A fee of transaction processing is given to Ledger manager a.

Get Operation. Get is an operation when State subscriber retrieves current cell state based on its requirement.

1. State subscriber u sends a ticket for querying a cell state $c_{x,y,t}$ to State broker b.
2. State broker b returns the cell state $c_{x,y,t}$ to State subscriber u. b continues the owner of $c_{x,y,t}$.
3. A fee of transaction processing is given to Ledger manager a.

4   Prototype System

To validate behavior of cell state exchange described in Sect. 3, we have been implementing a prototype system based on an existing open source block chain implementation.

Ethereum [4] was chosen in this research. The virtual currency in Ethereum is "ether". Ethereum provides not only money transfer functions, but also virtual machines for running user-defined program (smart contract), which is written in Solidity programming language. "Gas" is consumed to execute programs on third-party virtual machines in Ethereum. Since gas is regarded as a charge of program execution, owner of the program needs to pay "ether" according to consumption of "gas".

Based on the framework in Fig. 5, we have been implementing a prototype system (Fig. 6). The actors in Fig. 5 are implemented as smart contracts. They have Ethereum accounts, and they are owners of ether. State publisher registers a Provider contract, State broker registers a Broker contract, and State publisher registers a Publisher contract. These contracts can perform Publish/Subscribe/Get operations.

All data of occupancy grid map are maintained by a Broker contract. In the current implementation, cell states are represented by an array of pairs of cell state and timestamp.

5   Preliminary Experiment

A private block chain network is constructed in a private PC environment. The composition of PC used in the experiment is CPU Xeon E5-2665 2.40 GHz 2CRAM 32 GB, HDD 2 TB, Linux kernel 3.13.0-100 (Ubuntu 14.04). Ethereum environment in the experiment is Go Ethereum (geth) 1.7.3 [5], Mist Wallet 0.9.3 [6]. Synthetic data were generated for cell states of occupancy grid map.

5.1   Response Time

In the standalone environment, we measured response times for Publish operation and Get operation. In case of Publish operation, a response time means duration between the start time of Publish operation by Publisher contract and the receive time of payback notification from Broker contract. In case of Get operation, a response time means duration between the start time of Get operation by Subscriber contract and the arrival time of the requested cell state. In addition, as reference information, we measured response times for simple money transfer (User A sends 1 ether to User B). Note that every operation issues a new transaction, but the transaction has not yet been stored into a block (block creation time is shown in the next subsection).

Figure 7 shows the min/max/average of response times measured in 20 times. Simple money transfer takes 10 ms, Publish and Get operations, which need invocation of user-defined smart contract, take about 40 ms. We guess that the difference comes from overhead of invocations for smart contracts. Based on the result, we need to reduce the number of invocations for smart contracts. One of improvement ideas is passing multiple cell states at once.

5.2   Creation Time of the Block Containing the Transaction

Issued transactions are stored into a block, then the block is appended at the end of the block chain. Appending a new block takes several seconds, because an appropriate hash value for the block should be found.

Here, we measured creation time of the block for simple money transfer (User A sends 1 ether to User B). In Ethereum, a receipt is returned when the transaction is stored into a block. Thus, we measured receipt creation times. We notice that Ethereum dynamically changes difficulty of block creation. So, there may be difference between measurement in private block chain environment and one in global block chain environment.

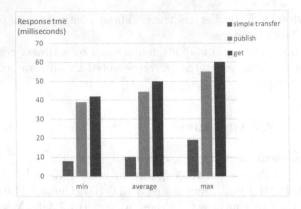

Fig. 7. Response time

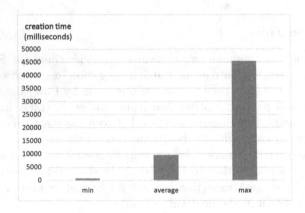

Fig. 8. Block creation time for simple money transfer

Figure 8 shows the min/max/average of block creation time measured in 20 times. Average time is about 10 s, and the worst time is about 45 s.

Based on this result, the system is not enough performance to maintain real-time data. It is difficult to apply for tracking moving objects. But, it is still applicable to share information about stable obstacles.

Based on the result of this preliminary experiment, we will improve the system.

6   Conclusion

This paper proposed a decentralized information sharing system for occupancy grid map by using block chain network. We presented the result of the preliminary experiment.

Future research issues include improvement of our system, and experiment in private cluster environment.

Acknowledgement. This research is partially supported by JSPS KAKENHI (JP16H01722).

References

1. Elfes, A.: Using occupancy grids for mobile robot perception and navigation. Computer **22**(6), 46–57 (1989)
2. Universal grid map library for mobile robotic mapping. https://github.com/ethz-asl/grid_map
3. Bitcoin Core. https://bitcoin.org/en/
4. Ethereum. https://www.ethereum.org/
5. Go Ethereum. https://geth.ethereum.org/
6. Ethereum Wallet and Mist. https://github.com/ethereum/mist

Location-Aware QoE-Driven Wireless Relay Deployment for Energy Efficient Multimedia Communications

Shuan He and Wei Wang$^{(\boxtimes)}$

Department of Computer Science, San Diego State University,
San Diego, CA 92182, USA
she@sdsu.edu, wwang@mail.sdsu.edu

Abstract. In order to support the future quickly deployable wireless multimedia services around mmWave frequency band, a location-aware multimedia relay node deployment (MRND) scheme is proposed in this paper. The self-organized MRND is capable of deploying the multimedia relay Base Station (BS) at the optimal locations in the network and apply optimal resource allocation strategy, to bridge the connections between the multimedia users and the internet gateway. The main contribution of this paper includes two folds. First, a Quality of Experience (QoE)-driven location-aware optimization algorithm of autonomous relay BS deployment scheme is proposed to achieve maximized system level QoE under energy constraints. Second, an adaptive resource allocation strategy is applied to adjust transmission data rate in conjunction to deployment optimization algorithm. Extensive simulation results in terms of the energy utilization and system QoE performance reveal the potential of the proposed self-positioning and rate allocation strategy in the MRNDs as a candidate for future multimedia communications in mmWave band.

Keywords: Location-aware relay deployment ·
Quality of Experience · Wireless multimedia communications

1 Introduction

Providing cost-efficient wireless multimedia internet services becomes a critical issue in extreme rural areas (e.g. deserts, very rural farms). Similarly, many one-time events or disaster events require a quickly deployment of a wireless networks to certain locations [1–4] providing multimedia services, for example, post-earthquake or flooding. The upcoming data services demanded by these communication systems have several challenges to tackle. Firstly, the large amount of multimedia (primarily video, voice, and images) data needs to be successfully collected and transferred to various residential locations. Moreover, different access technologies applied by different devices need to interoperate reliably between heterogeneous systems. What's more, the peculiar properties of the rural and

© Springer Nature Switzerland AG 2019
Y. Kawai et al. (Eds.): W2GIS 2019, LNCS 11474, pp. 160–173, 2019.
https://doi.org/10.1007/978-3-030-17246-6_13

outdoor environments increase the difficulty and costs of maintaining the batteries on-board. Finally, with the popularity of using the mmWave frequency band in communications, the users and the gateway may not have direct Line of Sight (LoS), thus promoting the utilization of relays in deploying multimedia services.

In this paper, the novel concept of location-aware Multimedia Relay Node Deployment (MRND) is proposed in order to support the intensive multimedia data communications in a quickly deployable setting around mmWave frequency band. The idea of MRND is to deploy a number of self-organizing mobile relay Base Stations (BS) at certain locations to maximize Quality of Experience (QoE) rather than throughput, in order to bridge the mmWave connections between users and the gateways. Figure 1 shows a typical MRND use case for post disaster mmWave multimedia communications in extreme environments. There are a large amount of users transmitting the information to the 5G/LTE or WiFi/Internet gateways through several mobile relay BSs. These relay BSs are connected to each other wirelessly with LoS forming the "relay bridge" to the backbone. Each relay BS covers a number of users with LoS too. The deployment of the mmWave relay BSs and the resource allocation at the relay BSs both have key impact of the network QoE and energy efficiency performance.

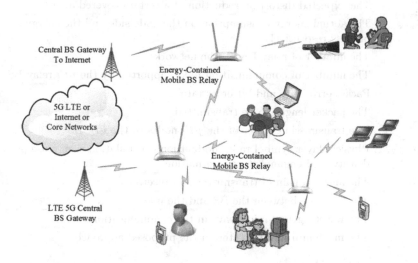

Fig. 1. Typical scenario of MRND around mmWave band.

We propose a new location-aware relay BS deployment scheme to achieve maximized system level QoE for multimedia users around mmWave frequency band. In this proposed approach, the system level QoE is modeled as total distortion reduction of multimedia, and the desirable BS location deployment is achieved by an optimization algorithm. Adaptive resource allocation strategy is also designed to adjust transmission data rate in conjunction to location deployment optimization algorithm.

The advantages of the proposed MRND approach are in three folds. First, it is QoE-driven rather than throughput-driven, and thus the positioning of relay BSs can satisfy the demands by the multimedia source users with various priorities. Second, the network resource is optimally relocated at the BSs so that the energy efficiency and QoE can be guaranteed. Third, it is self-healed in case of the failure of some relay BSs. With the flexible position and dynamic configuration of the BSs, the entire network improves the performance in an energy-efficient and QoE-optimized manner.

The rest of this paper is structured as followed. Section 2 discusses the related research in the literature. In Sect. 3, the location deployment problem is mathematically formulated. In Sect. 4, the analysis of location and strategy optimization is provided. In Sect. 5 the simulation results is shown and evaluated. The conclusion is drawn in Sect. 6. The symbols and notations in this paper are summarized in Table 1 for reading convenience.

Table 1. The notations used in this paper

Notation	Comments
r_i	The optimal resource allocation strategy of the i-th BS in the MRNDs
$E[D]$	The expected distortion reduction of a certain covered area
E, E_{max}	The actual energy consumption on the node side and the energy budget, respectively
N	The number of relay BSs in the networks
M_i	The number of communication links supported by the i-th relay BS
p, e	Packet error rate and bit error rate
L	The packet length to be transmitted
P_{ij}	The transmission power of the j-th nodes to the i-th BS
R_s, b, N_0, A	Physical layer symbol rate, modulation constellation size, noise power density, and channel state information
G_t, G_r	The antenna gain of transmitter and receiver
d	The distance between the BS and the node
λ	The wavelength of microwave in the communications
I_{max}	The maximum iteration step in the proposed approach

2   Related Work

Research on the location-aware energy efficient BS positioning in traditional throughput maximization networks has been studied in the past, but few discussed QoE-driven BS location deployment. In [5], the authors proposed an energy-efficient BS deployment scheme in cellular networks based on variable traffic process. The proposed scheme increases the energy utilization by choosing the optimal BS density according to the traffic load thresholds which switch off some BSs in the network. Recent studies include the pico-cell level BS deployment optimization. For example, the researchers in [6] studied the optimal deployment and power control scheme of the BSs in femtocells networks. However, such conventional fixed deployment schemes do not address the difficulties of the dynamically changing user demands and network resource constraints, with the increasing wireless nodes in today's networks. To tackle such challenges, many researchers have been focused to study the nodes deployment techniques in an autonomous self-organizing way that caters for the needs in rapid changing environments. The work in [7] is conducted to provide a comprehensive survey of various self-organizing and deployment in cellular networks. In [8], global and distributed algorithms for the mobile nodes deployment are studied. Research in [9] studies the strategy of sensor deployment by balancing the sensor density of a certain area. In [10], the connectivity and the distance between nodes are studied to reach reliable network performances. Another distributed scheme based on fluid dynamic models is proposed in [11]. The mobility of the sensor nodes achieves the optimal distribution by the self-spreading among the whole network. In [12], a two-step approach is proposed which reduces the nodes travel distance while maintaining their convergence property. Also, the problem of limited node communication range is overcome by improving the Lloyd's method. Similar algorithms for gateway self-deployment can be found in [13]. In [14], the mobility of BSs is discussed and a self-deployment and load balancing algorithm is proposed. However, the unequal importance and resource allocation of the multimedia communication are much ignored. Some research works evaluate the network resource management scheme to improve the performance and reliability in autonomous networks. The research in [15] studies interference management mechanisms in the autonomous Device-to-Device (D2D) network environments based on the optimal component carrier resource selection. A system level solution named autonomous network reconfiguration system is presented in [16] to solve the frequent failure problems in the wireless mesh networks.

Most previous work did not discuss the QoE-driven design of BS deployment. The joint consideration of BS deployment and multimedia network resource allocation has been overlooked. In this paper, we study the mobile relay BS deployment scheme with adaptive network parameter selection strategy, to achieve maximized system level QoE.

3 Problem Formulation

The objective of the MRND is to provide high-quality and reliable communication for large volume of multimedia transfer between various multimedia sources and the internet gateway at mmWave frequency band which requires LoS between transmitter and receiver. These devices should be organized and maintained autonomously and intelligently, since frequent manual intervenes are impractical due to the extreme physical conditions in disaster relief applications. The energy budget of the multimedia communication in MRND is restricted since the battery supply is not always guaranteed in the unstable environmental conditions. Therefore, new network deployment and configuration strategies should be introduced to: (1) provide the autonomous self-deploying and self-configuration network solutions; (2) ensure the energy budgets and QoE consideration among the nodes in the networks; (3) recover from emergency events such as link blackout or BS failure.

Two tiers of network architecture are adopted in order to reduce the burden of heavy multimedia communication system. The multimedia data from the source is transmitted back to the gateway via a mobile relay BS. Several mobile relay BSs operate in the open area at any position to support the communication links among a large number of sources in the MRND.

The objective of our work is to increase the multimedia transmission quality (e.g. distortion reduction) at system level in the MRNDs while keeping energy consumption budget of the nodes by optimizing the location of the relay BSs. The problem can be mathematically formulated as:

$$\biguplus\{(x_i, y_i), r_i\}_{i=\{1,\ldots,N\}} = \arg\max\{\mathbb{E}[\mathbb{D}]\} \tag{1}$$

subject to

$$E < E_{\max} \tag{2}$$

where (x_i, y_i) is the optimized position coordinate of the i-th BS with r_i as its the network resource parameter (e.g., rate, power, bandwidth, etc.), and $\mathbb{E}[\mathbb{D}]$ is the expected distortion reduction of a certain covered area. E and $E_{\max}$ denote the actual energy consumption and the energy budget, respectively.

4   Modeling and Analysis of Multimedia Relays

4.1   Mathematical Model for Distortion Reduction

In the MRND, the system level distortion reduction can be calculated as the summation of all the communication links connected to the relay BSs [17]. Let N denote the total number of BSs in the system, M_i denote the number of links supported by the i-th BS, $D_{i,j}$ and $p_{i,j}$ denote the distortion reduction and packet error rate respectively for the data transmission between the j-th node and the i-th BS, the expected QoE can then be expressed as:

$$\mathbb{E}[\mathbb{D}] = \sum_{i=0}^{N} \sum_{j=1}^{M_i} D_{i,j} * (1 - p_{i,j}) \tag{3}$$

Given the Bit Error Rate (BER) e and the packet length L, the packet error rate p is calculated as:

$$p = 1 - (1 - e)^L \tag{4}$$

Let P_t denote the transmission power, b denote the constellation size, R_s denote the source rate, N_0 denote the noise power density (w/Hz) and it's a preset system parameter. Let A denote the channel attenuation. Then the BER can be calculated as [18]:

$$P_t = \frac{2N_0 R_S(2^b - 1)}{3A} \left(erfc^{-1}\left(\frac{eb}{2\left(1 - \frac{1}{2^{\frac{b}{2}}}\right)} \right) \right)^2 \tag{5}$$

Assuming the channel attenuation A is modeled as the Free Space Loss (FSL), which is defined as the ratio between transmission power and receive power:

$$A = \frac{P_r}{P_t} = \frac{1}{G_t G_r}\left(\frac{\lambda}{4\pi d}\right)^2 \tag{6}$$

where λ is the signal wavelength, d is the distance between the receiver and the transmitter, G_r and G_t is the antenna gain of receiver and the transmitter respectively.

4.2   Energy Consumption Analysis

In the MRND, a variable size of communication nodes can be deployed randomly in the environments, so the network coverage and capacity can be flexible. The BS controls the start time and frequency allocation of each active communication link in a way similar to LTE networks. In such a design, the energy consumption can be modeled as the total communication energy consumed by all the communication links supported in the QoE coverage, which is shown as:

$$E = \sum_{i=1}^{N} \sum_{j=1}^{M_i} P_{ij} t + \sum_{i=1}^{N} P_{up} T_{up} \tag{7}$$

where P_{ij} is the transmission power of the j-th node to the i-th BS. $P_{up} T_{up}$ denotes the energy consumption of mobile BS for sending data to the central BS (gateway). Let L_{total} denote the data length to be transmitted over the link between the nodes and BS, so that the transmission time t can be calculated as: $t = L_{total}/(R_s b) + T_{overhead}$. Then it is substituted in Eq. 7, which is re-written as:

$$E = \sum_{i=1}^{N} \sum_{j=1}^{M_i} P_{ij}\left(\frac{L_{total}}{R_s b} + T_{overhead}\right) + \sum_{i=1}^{N} P_{up} T_{up} \tag{8}$$

This generic energy model can be modified to support other network protocol design incorporating the error-resilient schemes such as Automatic Repeat-reQuest (ARQ) or Forward Error Correction (FEC).

From Eqs. (3)–(8) we can see, the system QoE performance is directly related to the distortion reduction gain as well as the BER of the communication links. The distortion reduction gain is determined by the codec. The BER is affected by the transmission power and transmission distance. The network QoE can be improved by simply increasing the transmission power at each communication node side and decreasing the distance between the BS and the nodes. However, increasing the transmission power can also introduce a large amount of energy consumption and interferences if users are within a certain angle. In the disaster relief environments, the power supply is limited and will run out very quickly by choosing a larger transmission power constantly. Therefore, the trade-offs between the system performance and energy consumption needs to be carefully considered. With all the related factors discussed, we adopt an optimal BS position adaption strategy that the mobile BS works at the optimal location in the system to support the links for the covered nodes.

5   QoE Optimization and Rate Allocation

5.1   Optimization of QoE

The optimization problem can be solved by searching all possible positions in the network for BSs. Even though the centralized optimization provides the complete knowledge of the systems, the challenge of this approach comes from the computational complexity which exponentially grows with the number of communication nodes in the network. As pointed out in [14], similar problem to Eq. (2) is an NP-hard problem and thus cannot be simply solved by global search approach by the computational capability of BSs in a short time.

To overcome this challenge, a local optimization solution is proposed as Algorithm 1, which reduces the computational complexity by reducing the number and area of optimization process for each BS. In such distributed approach, we put forward the energy utilization ratio of the BS, which is used to maximize the best system performance while maintaining a low level of energy consumption. Let $(x_{opt}, y_{opt})_i$ denote the locally optimal position of the i-th BS, D_i and E_i denote the distortion reduction and the energy consumption of each link supported by the i-th BS respectively. Then the energy utilization is expressed as:

$$(x_{opt}y_{opt})_i = \arg\max \left\{ \frac{\sum_{j=1}^{M_i} D_j * (1 - p_j)}{\sum_{j=1}^{M_i} E_j} \right\} \tag{9}$$

This paper is focused on the system level QoE maximization under certain energy constraints by mobile BS positioning to adapt to the changing network environments. In another perspective, this objective of this problem is equivalent to maximize the system level energy utilization ratio (e.g., QoE performance per energy cost) that implies to support the communication links with minimum energy consumption. As discussed earlier, the high complexity and overhead of the global optimum solution is not a practical solution in the MRND, thus a distributed algorithm is discussed that requires low complexity but achieves close to the optimum solution.

Algorithm 1: Optimal Deployment of Relay BS Locations.

```
1. Obtain the initial information of the communication
   set for each BS.
2. For each communication set, optimize the location
```

```
     according to the largest energy utilization factor
     in Equation 9.
3.  At each BS, calculate the optimal rate allocation
     for each link in the current set according
     to Algorithm 2.
4.  Update the choice of relay BS for the source nodes
     (e.g. one can provide the largest transmission rate)
5.  After Imax iterations, terminate the optimizing process.
     Otherwise, repeat steps 1-4.
```

5.2   Rate Allocation Adaptation

To ensure the effectiveness of the proposed local optimization solution, automatic resource allocation scheme is applied based on the importance and distance of the communication nodes in the MRND. Such scheme can be achieved similar to the commonly used automatic modulation coding scheme in the WiFi or LTE networks. We assume that by the connections of every communication links, the BS is aware of the distortion reduction gain and distance of each node. So the rate allocation strategy can be applied at the BSs for source nodes with different importance and distance. Lower transmission rate is provided for the nodes with lower quality contribution and larger distance.

Even though in this paper, the transmission rate allocation strategy is discussed but other network resources such as the transmission power, FEC or ARQ schemes can also be applied in similar way. Let k denote the choice of the BS selected with the highest probable transmission rate $\mathbb{R}$, which can be expressed as:

$$k_{opt} \in [1, \ldots, N] = \arg\max(\mathbb{R} = R_s * b) \tag{10}$$

It is assumed that the multimedia source nodes connect to the relay BS that provides the highest transmission rate. In the proposed approach, the rate allocation is performed after the BS positioning local optimization algorithm. After updating its communication set that includes the nodes supported by each BS, another round of optimization is performed. During the BS hand over, the connections only change at the edges of each cell, so that such scheme archives the indirectly communication between each BSs and a global optimal effect.

Algorithm 2: Relay Rate Allocation in the MRND.

```
1.  Obtain the information of the current communication
     set for the BS.
2.  Calculate the channel conditions for all the
     communication links supported by the BS.
3.  Create an order list of all the supported nodes in
     the current communication set based on their channel
     conditions and the distortion reduction contribution.
4.  Allocate multi-rates adaptation schemes such as QPSK,
     16-QAM from the last node to the first node.
```

The concept of MRNDs is achieved by an iterative process that the position and resource allocation are optimized in each step at each BS. Therefore, the problem is solved by the proposed approach that includes two steps in each iteration of the optimization process: BS positioning and the rate control. The mobility BSs enables the MRND system to be self-organized without human intervene to reach the optimal positions. In addition, the resource management at the BS further contributes to the system QoE performance gain under limited energy budgets by exploring the source traffic and the channel conditions.

6   Simulation

In this section, we present the simulation results and analyze the performance of the proposed MRND. Several scenarios are simulated to represent different distribution of the source nodes. The results are presented with its solutions with the optimal positions and the system level energy utilization factor that indicates how much distortion reduction gain of the system under an energy consumption budget. The parameters in our simulation are presented as followed. For the multimedia data, we use the MEPG-4 encoded standard *Foreman* video sequence. The transmission power of the source nodes is 50 mW, the receive power of the BS is 500 mW, the receiver antenna gain is 10 dB at the BS, the transmitter antenna gain is 10 dB, the frequency used for communication is 60 GHz. The symbol rate is 2 MHz–20 MHz, widely used modulation schemes include QPSK and 16-QAM (b equals to 2 and 4 respectively). The number of mobile relay BSs N is 4 in the initial stage. The noise power density N_0 is -160 dBm/Hz. For fair comparison, the static BS deployment strategy where the BSs are uniformly distributed is selected to show the effectiveness of the proposed approach in the MRNDs. The maximum iteration step in the proposed approach is denoted as $I_{\max}$.

Figure 2 illustrates the uniformly distributed sources with equal distortion reduction gain deployed in the 100 m * 100 m area. Four mobile BSs start at randomly locations. After I_{max} rounds of optimization, four mobile BSs reach the optimal locations with optimal rate allocation strategy. From the figure we can see that after the completion of the optimization process, all four BSs reach the center of the corner areas of the larger square area on the planet surface, supporting the same number of source nodes. From the perspective of the source nodes, the BS with largest transmission rate is selected for the data transfer. The reason for such solution of this scenario is that equal importance in terms of distortion reduction and distances from BS result in the uniform distribution of BSs.

In the following up simulation shown as Fig. 3, a hot spot of six source nodes with higher distortion reduction (denoted as larger black solid circles in the figure) occur at the bottom left area in the same MRND map. To provide better

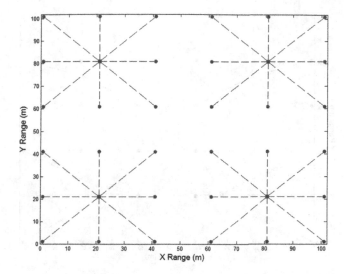

Fig. 2. Connections and optimal BS positions with equally distributed sources.

channel conditions to these nodes, the BSs move closer from previous center locations. Moreover, the rate allocation strategy during each optimization process allows the BS to assign higher rate transmission to the source nodes closer to the BS, which triggers the change of transmission rate at the border of four coverage area. Some nodes transfer the connection to a BS providing higher transmission rate, which in return triggers another new round of network optimization due to the update of the communication set. In this way, the MRND adaptively deploy the BS positions and organize rate allocation strategy to meet the dynamically changing nodes communication load. This figure demonstrates that the proposed MRND system has self-deploying characteristics that satisfies the future large amount of scientific data transfer on the outer planets.

Figure 4 shows the system distortion reduction performance in different node-BS distance scenarios. We take the Peak Signal to Noise Ratio (PSNR) as the measurement for multimedia distortion reduction (also QoE) in all our simulations. From this figure we can see that multimedia QoE is very sensitive to distance and transmission data rate. System would obtain high level of QoE when taking small data rate in long node-BS distance (e.g., 75 m). When the communication distance is short (i.e., 25 m), both low and high data rate have good distortion reduction performances in the MRND system. All these actions can be fully addressed in the proposed optimal dynamical resource allocation strategy.

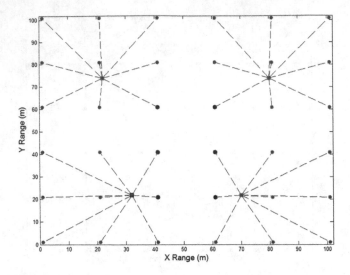

Fig. 3. Connections and optimal BS positions within a hot spot area with different unequal multimedia distortion reduction.

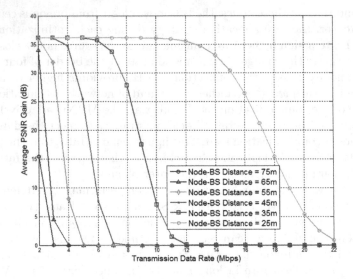

Fig. 4. System distortion reduction performances with different node distributions (in terms of distance), the QPSK modulation scheme is used in this simulation.

Then we explore the energy utilization factors in the proposed MRND with optimal BS deployment and static BS deployment strategies. As we can see from the simulation result shown in Fig. 5, energy utilization in the proposed optimal mobile relay BS deployment outperforms the static one in both QPSK and 16-QAM due to it's the dynamic resource (e.g., data rate) allocation features.

In addition, the overall energy utilization factor in QPSK modulation scheme (shown in the top subfigure) is lower than it in the 16-QAM in both optimal and static BS deployments. The rationale behind this is that the 16-QAM scheme has higher data transmission rate than QPSK, which causes less system energy consumption when transmitting the same amount of data.

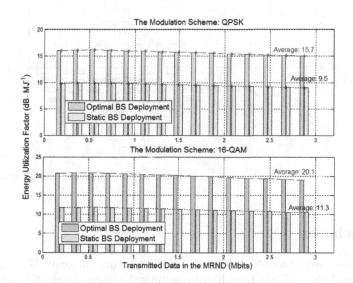

Fig. 5. Performance results of the energy utilization factor of the proposed approach versus that of the static BS deployment.

Besides the energy utilization performance in QPSK and 16-QAM modulations, we also investigate system QoE gain in both modulation schemes in Fig. 6. The amount of transmitting data is fixed in this simulation. We assume the nodes in the MRND have even distortion reduction and there are totally 4 mobile BSs serve them. As demonstrated in Fig. 6, the average PSNR gain from each node keeps linearly increase with the raising of energy budget in all transmission scenarios and the overall PSNR gain in QPSK (solid lines) outperforms the 16-QAM (dash lines) scheme. Recall that the 16-QAM modulation scheme has high energy utilization due to its high transmission rate. But on the other side, the BER in 16-QAM scheme is bigger than that in QPSK according to Eq. (5), which leads the overall QoE (PSNR gain) to go down. Furthermore, as shown in Fig. 6, benefiting from the high energy utilization in the front, the PSNR gain in 16-QAM scheme performs better before the energy budget of 40 J. But QPSK is the optimal modulation scheme in this case if energy is adequate.

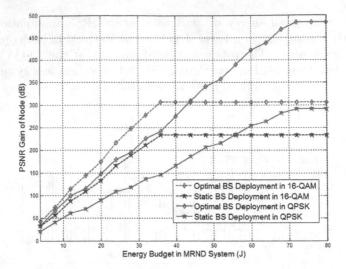

Fig. 6. System distortion reduction performances under different energy consumption budgets.

7   Conclusion

In this paper, we proposed a location-aware multimedia relay BS deployment system called the MRND to achieve maximized system level QoE while keeping to the energy consumption within the budget for mmWave frequency band wireless multimedia communications. An optimization algorithm of the desirable BS deployment was proposed to determine the locations, and an adaptive resource allocation strategy was designed to adjust transmission data rate in conjunction to the location-deployment optimization algorithm. Simulation results demonstrated that the proposed approach has improved the system QoE and the energy efficiency considerably while following the energy budget constraint in resource-constrained environments and scenarios.

Acknowledgments. This research was support in part by National Science Foundation Grant No. 1463768 on energy efficient wireless multimedia communications.

References

1. Sakano, T., et al.: Bringing movable and deployable networks to disaster areas: development and field test of MDRU. IEEE Netw. **30**, 86–91 (2016)
2. Miranda, K., Molinaro, A., Razafindralambo, T.: A survey on rapidly deployable solutions for post-disaster networks. IEEE Commun. Mag. **54**, 117–123 (2016)
3. Yarali, A., Ahsant, B., Rahman, S.: Wireless mesh networking: a key solution for emergency and rural applications. In: Second International Conference on Advances in Mesh Networks, Athens (2009)
4. Minh, Q.T., Shibata, Y., Borcea, C., Yamada, S.: On-site configuration of disaster recovery access networks made easy. Ad Hoc Netw. **40**, 46–60 (2016)

5. Wu, Y., Niu, Z.: Energy efficient base station deployment in green cellular networks with traffic variations. In: IEEE International Conference on Communications in China (ICCC), Beijing (2012)
6. Liu, J., Kou, T., Chen, Q., Sherali, H.D.: Femtocell base station deployment in commercial buildings: a global optimization approach. IEEE J. Sel. Areas Commun. **30**, 652–663 (2012)
7. Aliu, O.G., Imran, A., Imran, M.A., Evans, B.: A survey of self organisation in future cellular networks. IEEE Commun. Surv. Tutor. **15**, 336–361 (2013)
8. Heo, N., Varshney, P.K.: Energy-efficient deployment of intelligent mobile sensor networks. IEEE Trans. Syst. Man Cybern. Part A: Syst. Hum. **35**, 78–92 (2005)
9. Chang, R., Wang, S.: Self-deployment by density control in sensor networks. IEEE Trans. Veh. Technol. **57**, 1745–1755 (2008)
10. Li, J., Cui, L., Zhang, B.: Self-deployment by distance and orientation control for mobile sensor networks. In: International Conference on Networking, Sensing and Control (ICNSC), Chicago (2010)
11. Pac, M.R., Erkmen, A.M., Erkmen, I.: Scalable self-deployment of mobile sensor networks: a fluid dynamics approach. In: IEEE International Conference on Intelligent Robots and Systems, Beijing (2006)
12. Song, Y., Wang, B., Shi, Z., Pattipati, K., Gupta, S.: Distributed algorithms for energy-efficient even self-deployment in mobile sensor networks. IEEE Trans. Mob. Comput. **13**, 1035–1047 (2013)
13. Mullany, F.J., Ho, L.T.W., Samuel, L.G., Claussen, H.: Self-deployment, self-configuration:critical future paradigms for wireless access networks. In: Smirnov, M. (ed.) WAC 2004. LNCS, vol. 3457, pp. 58–68. Springer, Heidelberg (2005). https://doi.org/10.1007/11520184_5
14. Claussen, H.: Autonomous self-deployment of wireless access networks. Bell Labs Tech. J. **14**, 55–72 (2009)
15. Garcia, L.G.U., Pedersen, K.I., Mogensen, P.E.: Autonomous component carrier selection: interference management in local area environments for LTE-advanced. IEEE Commun. Mag. **47**, 110–116 (2009)
16. Kim, K.H., Shin, K.G.: Self-reconfigurable wireless mesh networks. IEEE/ACM Trans. Netw. **19**, 393–404 (2011)
17. He, S., Wang, W., Sohraby, K: Joint high level QP and low level power control in NOMA/OMA downlink wireless multimedia communications. In: IEEE Consumer Communications and Networking Conference, Las Vegas (2019)
18. He, S., Wang, W.: User-centric QoE-driven power and rate allocation for multimedia rebroadcasting in 5G wireless systems. In: Workshop on User-Centric Networking for 5G and Beyond IEEE Vehicular Technology Conference (VTC), Nanjiing (2016)

A Dynamic Weighted Trilateration Algorithm for Indoor Localization Using Dual-Band WiFi

S. Mathivannan, S. Srinath, R. Shashank, R. Aravindh,
and Vidhya Balasubramanian(✉)

Department of Computer Science and Engineering, Amrita School of Engineering,
Amrita Vishwa Vidyapeetham, Coimbatore, India
b_vidhya@cb.amrita.edu

Abstract. WiFi based indoor and semi-indoor localization techniques are essential components of indoor location-based services. Calibration-free techniques for WiFi signal strength based indoor localization can help make indoor localization systems scalable, cost-effective and easy to deploy. However, distance estimation errors and environmental factors affect the accuracy of non-calibration solutions like trilateration significantly, and addressing the accuracy issue is critical. To help improve accuracy, localization over dual-band WiFi (IEEE 802.11n) which uses both 2.4 GHz and 5 GHz bands is a potential alternative. This paper proposes a novel adaptive, weighted trilateration technique that uses the behavior of these two bands under different conditions. An iterative heuristic approach based on the characterization of the behavior of the bands is employed to determine the most likely position of a smart phone. Additionally optimization strategies are applied to improve the time complexity of this approach. Experiments conducted in different indoor environments show that our approach performs better than other non-calibration signal strength based approaches in terms of accuracy, and also reduces the worst case error.

Keywords: WiFi dual-band · Trilateration · Mathematical model

1 Introduction

Location-based services are a key component of Geographical Information Systems, and indoor location-based services are increasingly being used by application providers. Indoor localization and tracking play a major role in disaster response, navigation support, smart shopping services etc. However, there are still significant challenges to be addressed in order to make positioning practical in real indoor environments. These challenges include scalability, the trade-off

This work has been funded in part by DST (India) grant DyNo. C/4902/IFD/2016–2017.

© Springer Nature Switzerland AG 2019
Y. Kawai et al. (Eds.): W2GIS 2019, LNCS 11474, pp. 174–187, 2019.
https://doi.org/10.1007/978-3-030-17246-6_14

between cost and accuracy of the technology, and practical deployability of the localization system.

Multiple approaches have been employed for Indoor Positioning and some of the commonly used technologies are RF based systems (RFID (Radio Frequency Identification) [2,15], UWB (Ultra Wide Band) [5], WiFi [6]) and IMU based systems [1,4]. Among these, WiFi is the most preferred technology because, WiFi is cost effective [6,7] and most importantly has minimal infrastructural overhead. Additionally most buildings have WiFi infrastructure which makes localization using WiFi practical and with the proliferation of smart phones, WiFi based indoor localization systems are popular.

WiFi based indoor localization systems employ different parameters like Time of Flight, Time of Arrival, Angle of Arrival, Channel State Information [9,13] and Received Signal Strength Indicator (RSSI) [3]. The parameters like Time of Flight, Time of Arrival, provide accurate positioning but require specialized chipsets not commonly found in existing smart phones. Therefore RSSI is the most commonly used parameter and easy to access. However, the stability of RSSI is poor especially in indoor environments, therefore resulting in poor localization accuracy [3,10]. To address this, fingerprinting techniques have been proposed [6].

A fingerprinting based solution is generally achieved through two phases, viz., an offline tracking phase and an online tracking phase. During offline tracking, a comprehensive map of RSSI values in the indoor space is generated which acts as a training data. During the online tracking, the signal strength obtained during the run time from an unknown location, is matched to the closest possible value in the fingerprint map using kNN, Bayesian approaches, particle filter based approaches etc., to obtain the estimated location. Fingerprint techniques provide high accuracies [3], but are cumbersome to generate [11,12] and are not immune to dynamic changes in the environment. Fingerprinting techniques based on Channel State Information (CSI) propels the precision into the sub-meter level with several access points (AP) [13]. However, the performance is found to degrade with fewer APs mainly due to the limit of bandwidth. A rigorous and careful preprocessing becomes an important prerequisite for an effective implementation of fingerprinting, and hence not always practical.

Non-fingerprinting techniques like multilateration are easy to implement but are highly inaccurate. Several approaches have been proposed to reduce the cost of fingerprinting [17] or improve the accuracy of non-fingerprinting techniques like WALE, a weighted adaptive localization approach [10]. A less commonly used approach is to use dual-band based WiFi localization, which is the focus of this work. In recent times, IEEE 802.11n is gaining more attention due to the operation of devices in dual bands (2.4 GHz and 5 GHz) simultaneously. Earlier studies have shown that the 5 GHz band has higher localization accuracy than the 2.4 GHz band due to the stability and propagation effect exhibited by 5 GHz band [14]. This stability effect can suppress the various factors that hinder the accuracy of the distance estimated with log-distance path loss model. However, these properties are reliable only for ideal environments and there exists a need for studying and analyzing the performance of dual-band fused trilateration based localization in actual environments affected by the interference from weather, obstacles and human activities. Approaches to using dual-band for localization have primarily focused on fingerprinting solutions [8], and they have been found to increase localization accuracy.

It is imperative to design non-fingerprinting approaches that can combine the two bands, without the overhead of calibration. To do so requires effectively combining the advantages of 2.4 and 5 GHz bands so that algorithms like trilateration can be adapted suitably. One method is to adapt the trilateration to dual-band routers. However, errors in distance estimation affect the accuracy of the approach. Improving trilateration has usually relied on resizing circles and weighting or methods to improve router selection [10]. Resizing the circles helps improve accuracy but is not reliable when the distance estimation suffers highly from over or under-estimation errors. Usually, the direction and level of resize for each circle is not dynamically determined in existing methods. However a dynamic approach that customizes the resize required for each router based on both the bands can help in adapting to real-time conditions. In this paper we develop an integrated approach combining router selection, circle resizing and weighting and employs the properties of the 2.4 and 5 GHz bands for the same. We formulate the resizing problem as a minimization problem which aims to determine the best way in which the selected circles have to be resized and the level at which each of the circles has to be resized. The behavior of the two bands influences the properties of the circles from both the bands, and these properties are used as a weighting function, which is then applied in an iterative algorithm to estimate the best point of intersection based on above criteria which is the maximum likelihood position. By employing this integrated approach we aim to improve the indoor positioning accuracy with minimum overhead. The specific contributions of our work are as follows:

- Formulating the trilateration problem as a minimization problem which aims to determine the best resizing of the circles, both in terms of direction and distance.
- An intelligent router selection method that employs the properties of the two bands.
- Design a Dynamic Weighted Dual-band assisted Trilateration algorithm for indoor spaces incorporating an optimal iterative approach solving the above minimization problem.

In the next section, we formulate the trilateration minimization problem and provide an approach to solving it. Next, we explain in detail the overall approach designed to obtain the maximum likelihood position of a given candidate point in a given test area incorporating the solution to the minimization problem. Section 4 describes the experimental setup and compares our approach to standard non-fingerprinting approaches involving trilateration. Finally, we conclude with observations in Sect. 5.

2   Formulating Trilateration as a Minimization Problem

Trilateration in an ideal scenario results in three circles intersecting at a point, this point is considered as the original position of the object. But in real life scenarios, we find that this does not happen, and most of the times the distances are overestimated or underestimated, resulting in circles not intersecting or being too large with large overlapping regions. To counter this problem, various algorithms have been proposed which involves resizing the circle based on whether they are overestimation or underestimation like WALE [10], DCE [16] etc. We address this issue by formulating it as a minimization problem, which is solved mathematically to get the best likelihood position.

Consider three routers chosen for trilateration, whose RSS values from the localization device are $R1, R2$, and $R3$ respectively, and the distance estimated using the log distance path loss model is correspondingly r_1, r_2, and r_3, which form the radii of the three circles. x_i, y_i being the centers of these circles. Let w_1, w_2 and w_3 be the weights assigned to the three circles so that the resizing of the circles are not uniform. The weights are derived from a set of observations based on experiments in real environments and certain established theoretical knowledge, which will be discussed subsequently.

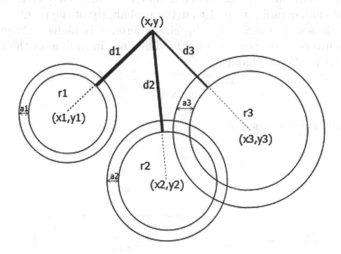

Fig. 1. Estimating the candidate point

Figure 1 shows three circles and their estimated radii. Our goal is to find the candidate point x, y whose weighted distances to the circumference of the three circles is minimum. Let these distances be d_1, d_2 and d_3 respectively. From the figure, di can be calculated as

$$d_i = \sqrt{(x_i - x)^2 + (y_i - y)^2} - r_i$$

where $r'_i = d_i + r_i$ is the resized circle radius. The problem is now formulated as follows: To find the candidate point x, y such that

$$[d_1^2 * w_1 + d_2^2 * w_2 + d_3^2 * w_3] \text{ is minimized.}$$

To reduce the impact of errors introduced when the estimated point is much closer to one router, we employ a squared dependency instead of a linear dependency to d_i.

Next we derive the weighting function based on the observations made during our experiments. The value of wi depends on the following factors.

Circle Radius: The smaller the radius of the circle, the higher the probability of the candidate point lying near the circumference of the circle (since Wi-Fi signals are known to deteriorate with distance, it becomes more unpredictable as radius increases). This has been observed in prior studies as well [10]. Hence smaller circles are given more weightage.

$$w_i \propto \frac{1}{r_i^2}$$

Difference in Radii of 2.4 GHz and 5 GHz Circles: For each router two circles are formed, one for the 2.4 GHz and the other for 5 GHz. If the two circles are of similar radii, we find that the probability of the point being inside both these circles, and closer to the circumference is higher. Essentially the weighting is inversely proportional to the difference in radii a_i of the two circles corresponding to the two bands.

$$w_i \propto \frac{1}{a_i}$$

Combining these two we get,

$$w_i = \frac{1}{r_i^2 * a_i}$$

Therefore,

$$P_{x,y} = \min \left[\sum_{i=1}^{3} \frac{\left(\sqrt{(x_i - x)^2 + (y_i - y)^2} - r_i \right)^2}{r_i^2 * a_i} \right]$$

The minimization problem incorporating this weighting function is then solved to find the estimated location. The next section will discuss the details of our algorithm that estimates the indoor position of a mobile device.

3   Dynamic Weighted Dual-Band Assisted Trilateration Algorithm

This section describes our overall algorithm, which estimates the likelihood position using the weighting function defined in the previous section. Initially, on obtaining a set of dual-band RSSI readings from any candidate point in a given area from all the APs in range, a preprocessing of those readings is carried out. The preprocessing involves filtering of outliers that hinder the stability of rest of the readings, and distance estimation using the most commonly used log-distance path loss model, after which a systematic case wise analysis of dual band is performed to shortlist either of the circles from a pair of circles formed by the dual-band in each of the routers. This is then followed by the selection of three-nearest routers based on the values of the distances obtained. The minimization problem is formulated for the circles that were shortlisted from the previous step. The solution to this trilateration minimization problem is considered as the maximum likelihood position of the candidate point.

Initially for any candidate point in a given test area, RSSI values of both 2.4 and 5 GHz are collected periodically from a set of N dual-band routers. From each band of a router, a sample set of RSSI readings are collected from the candidate point and are processed further for distance estimation. Generally, the mean of these values are considered for normalizing the values to a single consolidated reading for the distance estimation. However, the mean could be erroneous as it is prone to outliers. Therefore we estimate the median value, which has been found to be a more reliable estimate of the RSSI. The distance is estimated using the standard log- distance path loss model as mentioned earlier [18], where each of the RSSI value is converted to their respective distances. The log-distance path loss model formula is as follows.

$$D = 10^{\frac{P - P_0}{10*N}}$$

where D is the estimated distance, P is the obtained RSSI, P_0 is the RSSI reading at 1 m from the router and N is the normalizing constant which depends upon the environment.

3.1   Router Selection Algorithm

For each of the routers, we now have two circles from both 2.4 GHz and 5 GHz. For our Weighted Dual Band trilateration, we only need one circle corresponding to each router. For determining the circle to be chosen, we propose a decision algorithm that exploits the properties of 2.4 GHz and 5 GHz. We know that 5 GHz performs well for shorter distances due to its stability and the propagation effect attributing to its higher frequency [8]. For longer distances, we observe that the 2.4 GHz performs much better than the 5 GHz due to the higher wavelength of the former.

A simple calibration experiment has been conducted to understand the behavior of both the bands. In the given environment, we place a router, and measure the RSSI at intervals of 1 m from the router. The distance is estimated using the log-distance path loss model for both the bands and the obtained error in distance is plotted as shown in Figs. 2 and 3.

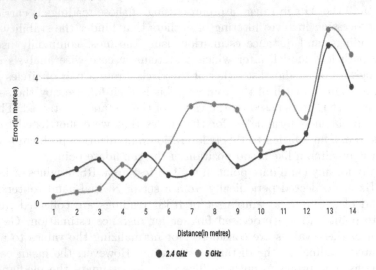

Fig. 2. Distance vs Error (Indoor)

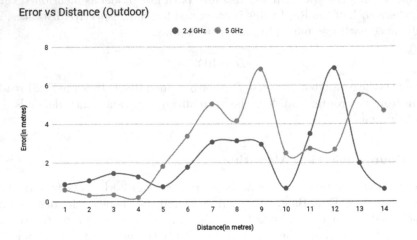

Fig. 3. Distance vs Error (Outdoor)

From the graphs in Figs. 2 and 3, it is observed that for smaller distances 5 GHz perform better in general, for longer distances 2.4 GHz performs better and for distances in between, either of them can show higher accuracy. The stability of the 5 GHz band, its poor performance around obstacles, and the penetration capacity of 2.4 GHz band leads to this performance variability. Let α be the threshold distance below which the 5 GHz band's accuracy is clearly higher, and β be the distance above which the 2.4 GHz performs better. When distance is less than α, preference is given to the 5 GHz band circle, and when distance is greater than β the algorithm chooses the 2.4 GHz band circle. For the distances between α and β, where there is no clear indication of the better performing band, the smaller of the two circles is chosen. Let the radius of the 2.4 GHz circle be called as $r_{2.4}$ and radius of 5 GHz circle be called as r_5. The band selection algorithm that uses the above observation works as shown in Table 1.

Table 1. Band selection

2.4 GHz band	5 GHz band	Chosen band
$r_{2.4} < \alpha$	$r_5 < \alpha$	5 GHz
$r_{2.4} < \alpha$	$\alpha < r_5 < \beta$	5 GHz
$r_{2.4} < \alpha$	$\beta < r_5$	2.4 GHz
$\alpha < r_{2.4} < \beta$	$r_5 < \alpha$	5 GHz
$\alpha < r_{2.4} < \beta$	$\alpha < r_5 < \beta$	Smaller of the two
$\alpha < r_{2.4} < \beta$	$\beta < r_5$	2.4 GHz
$\beta < r_{2.4}$	$r_5 < \alpha$	5 GHz
$\beta < r_{2.4}$	$\alpha < r_5 < \beta$	2.4 GHz
$\beta < r_{2.4}$	$\beta < r_5$	2.4 GHz

For effective band selection, it is important to determine the threshold values of α and β, which depend on the environment of interest. The value of α and β was calculated based on the calibration experiment described earlier (see Figs. 2 and 3. For our indoor environment, we find that $\alpha = 3$ and $\beta = 6$ gives the best results. Similarly, for semi-indoor environment, we observe that $\alpha = 4$ and $\beta = 12$ gives the best results. After arriving at a single band for each router, we now choose the nearest three routers with the minimum radius. The optimization function for resizing these circles is then applied.

3.2   Weighted Dual Band Trilateration

In order to estimate the likelihood position based on the chosen three circles, the given test bed is divided into smaller square grids of length δ, and for each of the cells in the given area, the likelihood of the point lying in that grid is computed. Let the length of the test area be 'L' meters and breadth be 'B' meters. The proposed algorithm is a greedy iterative approach that iterates over each cell and computes likelihood $P_{x,y}$ of the actual point x, y lying in that grid as shown in Algorithm 1.

Algorithm 1. Dual-band Weighted Trilateration

---

for $x \leftarrow 0, x < L, x \leftarrow x + \delta$ **do**
    for $y \leftarrow 0, y < B, j \leftarrow y + \delta$ **do**

$$P_{x,y} = \min \left[\sum_{i=1}^{3} \frac{\left(\sqrt{(x_i - x)^2 + (y_i - y)^2} - r_i \right)^2}{a_i * r_i^2} \right]$$

    end for
end for

It is evident from the algorithm that as the δ decreases, the number of iterations for a given test area increases. We improve the efficiency of the algorithm and reduce its time complexity by introducing the following iterative pruning approach.

Initially, the entire test region is cropped to a smaller rectangular area, whose boundaries are such that all three chosen circles are enclosed within the rectangle. This is virtually treated as a grid of equally spaced squares. The intersection points in the grid are taken to be the iterative points. On applying the algorithm to these intersection points in the grid, the point with maximum likelihood probability is marked for the next iteration. Now, the four square regions enclosing that point is further subdivided into a smaller sized square grid whereas other square regions in the grid can be pruned. However, there exist few edge cases, which should be treated a little differently as follows.

– For corner points, only one region has to be considered for the next iteration because there exists only one region in the corner surrounding it.
– For the points in the edges of the rectangle, two regions should be shortlisted for the next iteration because there will be two neighboring square regions to that point.

The intersection points from the newly obtained grid are taken as candidate points for our algorithm. Similarly, repeating this process with finer grids on selected regions until a required δ gives the maximum likelihood position of the candidate point in the final iteration. Clearly, this avoids most of the unnecessary computations during the weighted dual band trilateration algorithm. To explain the technique with more details we take an example of a rectangular environment.

Consider a rectangular environment, as shown in Fig. 4, with L x B dimensions. Dividing this environment's length and breadth with $\delta 1$ results in a grid, **G**, with equally spaced, square regions each of side length $\delta 1$. Now, each intersection point in the grid **G** is an iterative point, which on application of the weighted dual band trilateration algorithm for **G** results in an intersection point with maximum likelihood score. Consider this point to be **P1** as mentioned in the diagram. The neighbouring regions around the point **P1** are **R6**, **R7**, **R10** and **R11**. These regions together form a finer grid, **G1**, whose length and breadth has been subdivided with $\delta 2$. On iterating the intersection points from **G1** with

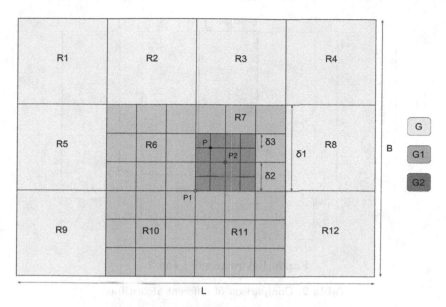

Fig. 4. Greedy strategy for the selection of iterating point in the grid

the algorithm results in another point with the maximum likelihood score, say, **P2**. On repeating this process with finer grid **Gn**, the final maximum likelihood point, **P** is obtained. In our experiment, the initial δ value was taken to be 3 meters and we carried out 3 iterations (**G3**) where $\delta 3 = 0.1$ m for arriving at the result of maximum likelihood position.

We observe that this pruning approach helps remove the overhead in search, and reduce the complexity of the algorithm, thus making the algorithm efficient in resource constrained devices, while not compromising on accuracy.

4 Experimentation and Results

In order to experiment, and analyze the performance of our proposed algorithm, two environments were selected. The first environment, described in Fig. 5 is a 35 m × 9 m space that is a typical office environment which is a completely closed indoor space with multiple cubicles and glassed rooms. The second environment (41 m × 38 m), illustrated in the same Fig. 5, includes the first environment and is a semi-indoor space that contains a partial outdoor space inside the building. The experiments were conducted with minimal to moderate human activity. For the experiments, we used 4 and 12 dual-band routers for environment 1 and 2 respectively. The goal was to use as minimal number of routers as possible.

From our experiments, we observed the properties of the circles formed by the dual-band. Based on the multiple circle patterns and our experiences from these experiments, we assume a set of router positioning conventions that may be followed to minimize the experimental errors. Our approach works even if the below mentioned conventions are not followed.

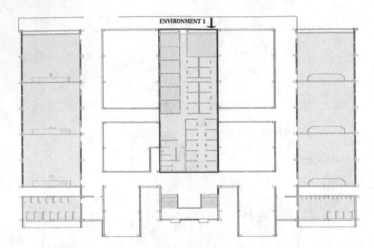

Fig. 5. Environments 1 and 2

Table 2. Comparison of different algorithms

		Our approach	Normal trilateration	Cyclic increase
E1	Mean error (in m)	2.54	4.47	3.98
	Stdev	1.41	3.02	2.28
E2	Mean error (in m)	2.408	4.94	3.15
	Stdev	1.647	3.01	1.98
E3	Mean error (in m)	4.087	12.92	7.64
	Stdev	2.39	6.02	4.11

1. Routers are placed such that they are not collinear, and are at the same height
2. Routers are also positioned in such a way that the neighbouring ones are equidistant from each other.

Three sets of experiments were conducted. The first experiment (E1) was conducted using the fingerprinting dataset over the environment 1 obtained from [8], which included 4 routers. The next experiment (E2) involved readings obtained in environment 1 with 6 routers. The last experiment (E3) was conducted in environment 2 which contains 12 routers, and readings were obtained primarily from the semi-indoor regions. To compare our algorithm, we choose the basic trilateration in 2.4 GHz, and the trilateration with cyclic increase algorithm part in [10]. All three algorithms have similar preprocessing to remove outliers in the RSSI readings.

The results of our experiments as shown in Table 2 indicate that our approach performs much better than the normal trilateration and the trilateration after cyclic increase. There are two reasons for this performance. The router selection in our approach helps in choosing the best routers so that erroneous values can be

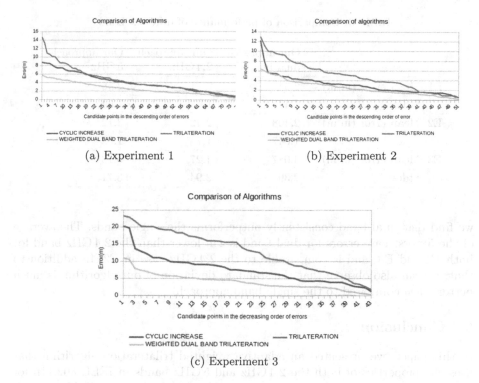

Fig. 6. Error plots of dual-band weighted trilateration vs other approaches

removed. Secondly, the iterative greedy approach helps in choosing the optimal position while considering the benefits of both the bands. The primary contributor to the accuracy here is the position estimation method. It helps improve the accuracy consistently in different types of environments, with the minimum number of routers. Additionally, the low standard deviation is an indicator of the stability of the approach. Figure 6 shows the improved performance even in the worst case when using dual-band approach.

The primary reason for proposing a dual band algorithm is to overcome the pitfalls of a single band algorithm. We now evaluate our algorithm further by running our algorithm with both the bands and single bands. Table 3 outlines the performances of our algorithm when run using only 2.4 GHz, only 5 GHz and when both the bands are used. From Table 3 you can see that our algorithm performs comparable to the dominant band in that environment and better than that. However the performance is not significantly better than the dominant band since, the environment was skewed towards 2.4 GHz based on our analysis. More specifically, the environment had more points with better accurate readings for that band. In general, in the presence of obstacles and clutter like cubicles the 2.4 GHz band performs better, while the 5 GHz band performs better in clearer spaces. On closer inspection, we observed that the worst case performance improves when both bands are used. On observing the top 5 worst case errors,

Table 3. Comparison of performance of different bands

		Our approach dual band	Our approach 2.4 GHz	Our approach 5 GHz
E1	Mean error (in m)	2.54	2.71	2.56
	Stdev	1.42	1.63	1.54
E2	Mean error (in m)	2.408	2.42	3.12
	Stdev	1.647	1.7	1.89
E3	Mean error (in m)	4.087	4.27	5.13
	Stdev	2.39	2.94	3.71

we find that dual band consistently outperforms the single bands. The average of the 5 worst case errors for dual band is 2 m lower than the 2.4 GHz band for both E1 and E3, and is comparable to the 2.4 GHz case in E2. In addition to that, we can also observe that the standard deviation of our algorithm is much better when compared to the single band approach.

5   Conclusion

In this paper, we presented an adaptive weighted trilateration algorithm that uses the properties of both the 2.4 GHz and 5 GHz bands in IEEE 802.11n for indoor localization. We formulated the problem of finding the likelihood estimation as an optimization problem that minimizes the weighted distance of the point to three circles. We also proposed a novel router selection algorithm that incorporates the properties of the circles generated in both bands. Our experiments over different datasets obtained from both indoor and semi-indoor environments show that our approach outperforms current non-fingerprinting techniques using RSSI and is also stable. Further analysis has revealed that using both the bands helps improve the accuracy and stability over single band based localization. In environments that are larger with a combination of open spaces and clutter, we expect a significant improvement in accuracy since both bands play a role. Future work will involve identifying trilaterability of points and reducing errors further. The robustness of our approach, easy deployability, and accuracy make our algorithm suitable for use in indoor location-based services and adaptation of these services more widely in indoor spaces.

References

1. Ibrahima, M., Moselhib, O.: IMU-based indoor localization for construction applications ISARC. In: Proceedings of the International Symposium on Automation and Robotics in Construction (2015)
2. Bilodeau, J.-S., Bouzouane, A., Bouchard, B., Gaboury, S.: An experimental comparative study of RSSI-based positioning algorithms for passive RFID localization in smart environments. J. Ambient. Intell. Hum. Comput. **9**, 1–17 (2017). https://doi.org/10.1007/s12652-017-0531-3

3. Wen, Y., Tian, X., Wang, X., Lu, S.: Fundamental limits of RSS fingerprinting based indoor localization. In: 2015 IEEE Conference on Computer Communications (INFOCOM), Kowloon, pp. 2479–2487 (2015)
4. Yang, Z., Wu, C., Zhou, Z., Zhang, X., Wang, X., Liu, Y.: Mobility increases localizability: a survey on wireless indoor localization using inertial sensors. ACM Comput. Surv. **47**(3), Article 54 (2015)
5. Shi, G., Ming, Y.: Survey of indoor positioning systems based on ultra-wideband (UWB) technology. In: Zeng, Q.-A. (ed.) Wireless Communications, Networking and Applications. LNEE, vol. 348, pp. 1269–1278. Springer, New Delhi (2016). https://doi.org/10.1007/978-81-322-2580-5_115
6. Xiao, J., Zhou, Z., Yi, Y., Ni, L.M.: A survey on wireless indoor localization from the device perspective. ACM Comput. Surv. **49**(2), Article 25, 31 p. (2016)
7. Gu, Y., Lo, A., Niemegeers, I.: A survey of indoor positioning systems for wireless personal networks. IEEE Commun. Surv. Tutor. **11**(1), 1332 (2009). https://doi.org/10.1109/SURV.2009.090103
8. Shanmugaapriyan, P., Chitra, H., Aiswarya, E., Balasubramanian, V., Kumar, S.A.: A pragmatic approach for effective indoor localization using IEEE 802.11n. In: Garcia Pineda, M., Lloret, J., Papavassiliou, S., Ruehrup, S., Westphall, C. (eds.) Ad-hoc Networks and Wireless. ADHOC-NOW (2014)
9. Xiaozhen, Y., et al.: IT-QEAS: an improved trilateration localization method through quality evaluation and adaptvie optimization selection strategy. In: 2017 Prognostics and System Health Management Conference (PHM-Harbin), Harbin, pp. 1–6 (2017)
10. Sundar, D., Sendil, S., Subramanian, V., et al.: WALE: a weighted adaptive location estimation algorithm. Ambient Intell Human Comput (2018). https://doi.org/10.1007/s12652-018-0940-y
11. Abu-Sharkh, O.M.F., Al-hamad, A.M., Abdelrahim, T.M., Akour, M.H.: Dynamic multi-band allocation scheme for a stand-alone wireless access point. In: 2012 26th Biennial Symposium on Communications (QBSC), pp. 168–173, May 2012
12. Liu, H., Darabi, H., Banerjee, P., Liu, J.: Survey of wireless indoor positioning techniques and systems. IEEE Trans. Syst. Man Cybern. Part C Appl. Rev. **37**(6), 10671080 (2007)
13. Chen, C., Chen, Y., Lai, H., Han, Y., Liu, K.J.R.: High accuracy indoor localization: a WiFi-based approach. In: 2016 IEEE International Conference on Acoustics, Speech and Signal Processing (ICASSP), Shanghai, pp. 6245–6249 (2016)
14. Parvin, H., Alizadeh, H., Minati, B.: A modification on k-nearest neighbor classifier. Glob. J. Comput. Sci. Technol. https://computerresearch.org/index.php/computer/article/view/650
15. Bahl, P., Padmanabhan, V.N.: RADAR: an in-building RF-based user location and tracking system. In: Proceedings of 19th Annual Joint Conference of the IEEE Computer and Communications Societies (INFOCOM 2000), Tel Aviv, Israel, vol. 2, pp. 775–784, March 2000
16. Jiang, J.A., et al.: A distributed rss-based localization using a dynamic circle expanding mechanism. IEEE Sens. J. **13**(10), 3754–3766 (2013)
17. He, S., Ji, B., Chan, S.G.: Chameleon: survey-free updating of a fingerprint database for indoor localization. IEEE Pervasive Comput. **15**(4), 66–75 (2016)
18. Soorty, B., Sarkar, N.I.: Finding a closest match between wi-fi propagation measurements and models. In: 2015 2nd Asia-Pacific World Congress on Computer Science and Engineering (APWC on CSE), pp. 1–6. IEEE (2015)

Author Index

Printed in the United States
by Bookmasters

Printed in the United States
By Bookmasters